Just Culture and the Criminalisation of Air Accidents

Just Culture and the Criminalisation of Air Accidents equips the reader with the tools to understand the meaning of criminalisation in the civil aviation industry. It enables them to resolve problems that arise in risk management within the context of their professional employment.

By examining the priorities of pilots, air traffic controllers and prosecutors, this book analyses the anatomy of criminalisation in civil aviation and works towards constructing solutions that protect the professionals' human rights and the Just Culture concept within the framework of criminal law and practice. It covers the duty of an air traffic controller to take the standard of care to guard against acts or omissions which might cause loss and give rise to a claim in negligence. The book illustrates the critical features discussed with numerous case studies from States and jurisdictions around the world.

The book will benefit airline operators, pilots, air traffic controllers or managers, and aviation business managers.

Just Culture and the Criminalisation of Air Accidents

Simon Daniels

CRC Press
Taylor & Francis Group
Boca Raton London New York

CRC Press is an imprint of the
Taylor & Francis Group, an **informa** business

Designed cover image: Shutterstock

First edition published 2025
by CRC Press
2385 NW Executive Center Drive, Suite 320, Boca Raton FL 33431

and by CRC Press
4 Park Square, Milton Park, Abingdon, Oxon, OX14 4RN

CRC Press is an imprint of Taylor & Francis Group, LLC

ISBN: 978-1-032-78910-1 (hbk)
ISBN: 978-1-032-78911-8 (pbk)
ISBN: 978-1-003-48969-6 (ebk)

DOI: 10.1201/9781003489696

Typeset in Times
by codeMantra

Contents

About the Author

Simon Daniels qualified as a lawyer in 1984. Since then, he has specialised in criminal and marine litigation, before developing a career in alternative dispute resolution as a qualified dispute resolver. He brought his professional background to the academic field at Solent University, where he managed the law and business parts of the MCA-accredited Merchant Vessel Operations programme at Warsash Maritime Academy, one of the world's leading centres of excellence in training for merchant shipping professionals. While at Solent, he was involved in a consultancy capacity with specialist advice in a number of criminal cases, including the trial of a Watchkeeper for manslaughter and a jet ski rider under s58 Merchant Shipping Act 1995. In 2012, Simon was awarded a PhD for his thesis: The Criminalisation of the Ship's Master. In 2014, he briefed an international conference hosted by Eurocontrol and the Dutch government on criminal negligence in matters arising out of the MH17 incident. He has published several academic papers and, in 2016, wrote a book titled *Corporate Manslaughter in the Maritime and Aviation Industries*. More recently, he has written a number of books on criminalisation and risk management in merchant shipping, all published by Routledge, and leads courses in criminalisation for the Nautical Institute.

INTRODUCTION – AIMS AND OBJECTIVES

Don't bring me a problem without offering me a solution.

This was a mantra which the author used frequently when in private practice as a lawyer; and that is what we have to do with this book.

The Pilot-in-command has the ultimate authority for their aircraft and retains absolute discretion, remaining accountable for their command decisions. But the Pilot had better have good evidence that they managed the risk to the standard demanded by an evolving area of the law.

Equally the air traffic controller has a duty to take the standard of care, reasonably expected of such a professional in their position, to guard against acts or omissions which might cause loss and give rise to a claim that the Court would find so serious as to make it a crime.

The scenario which makes criminals of professionals who were merely negligent had once been unknown, yet the criminalisation process is growing and creating dangerous precedents, undermining the very purpose of the Eurocontrol model policy's Just Culture, to promote continuous learning from previous mistakes and to encourage professionals to trust in the process without fear of punishment for the human element of errors. The dividing line between an error of judgment and gross negligence is notoriously thin and ill-defined, though, leading to the fearful comment, "There but for the grace of God go I." But the author was alarmed by the astute observation of Michaelides-Mateou and Mateou[1] that aviation professionals have little knowledge and understanding of the law which draws this dividing line, which makes this book essential reading.

That is at the core of the criminalisation problem, and demands a solution that both the prosecutor and the individual can live with. Failure to find a solution is unthinkable; but that is what Chapter 7 is all about.

The aim of this book is to equip aviation professionals with the knowledge that enables them to join the informed debate around criminal accountability in civil aviation. In this book, the author will analyse the anatomy of criminalisation, examining the priorities of the professionals and those of the prosecutors, in order to get a clear idea of the conflict, and how that conflict can be resolved.

So this book has clear objectives to enable aviation professionals to make up their own minds about the problems and solutions with criminalisation. Those objectives have to be met in this order:

1. Synthesise and appraise information arising out of the conduct of prosecuting authorities and apply it effectively to the legal theory which distinguishes an error of judgment from criminal negligence. In this way, the author establish a foundation of the critical areas of law and practice upon which criminalisation is founded.
2. Having addressed this in the early chapters, the author then develop our knowledge and understanding of the theory of Just Culture. Armed with this knowledge, the author can then generate persuasive conclusions to the complex problems arising out of the clash between criminalisation and Just Culture, with a view to analysing how the Just Culture embraced in international law is working in relation to safety and justice, within the context of the human element. This objective is not entirely linear, though we need to keep an open mind about our understanding process, as if it were slicing into a layer cake: each new layer builds on the previous one.
3. Building on our knowledge and understanding in this way, we can then exercise informed critical judgment to reflect upon and evaluate the options for solutions in criminal law and practice, restoring trust against injustice, in order to support the implementation and dissemination of a Just Culture concept.
4. By addressing these objectives, we can equip ourselves with the knowledge and skills to make up our own minds about where the criminalisation process is taking aviation professionals, whether we like it or not; and from there, we can form our own conclusions to the solutions in Chapter 7, by asking one of two questions: *either*, Is that right? *Or*, Surely that can't be right? It is up to each of us to form our own critical judgment.

NOTE

1 Michaelides-Mateou S and A Mateou, 2010, *Flying in the Face of Criminalization: The Safety Implications of Prosecuting Aviation Professionals for Accidents*, Routledge, Abingdon. And see Chapter 2 below.

1 Crime and Punishment

OVERVIEW

Crimes are wrongs which threaten the wellbeing of Society and, to that extent, they go beyond the mere interference with some individual's private right; in short, Society must be protected against such wrongs.

So a crime is whatever Society says it is. People elect their lawmakers on the promises that are made to them, and it is this which underpins the concept of Justice, upon which laws are made and Prosecutions conducted. It is the evolutionary process of criminal accountability, defined and applied by rules that differ according to the normative ethics of changing societies, which has forged the process which the global aviation community identifies as the phenomenon of criminalisation. It has developed not by the intent, but by the harm caused, and certainly should only be applied when it is fair to do so.

What is fair? Fairness demands equal rights for all individuals, as a principle of justice which must be satisfied before other political interests are satisfied. But the controversy surrounding the criminalisation of professionals in the aviation industry is founded on the perception that they are burdened with criminal accountability beyond that accorded to any other private citizen, by virtue of their risk management functions. In this chapter, we achieve an understanding of how criminal liability works, giving us a picture of the basic characteristics of a crime.

Recent developments in legislation add another layer to the question of fairness, in statutory wrongs that dispense with the need for proof of a guilty mind, and focus on the State's role of punishing the offender without the need for proving all the elements of a crime. Perhaps most seriously, in the UK's Civil Aviation Act 2012, the Civil Aviation Authority (CAA) has been given powers to make use of alternative civil sanctions alongside existing criminal penalties; given the distinction between criminal and civil law, this must increase the risk of ever-wider criminalisation.

What is a crime? Crimes are wrongs which threaten the well-being of Society and, to that extent, they go beyond the mere interference with some individual's private right; in short, Society must be protected against such wrongs. Smith and Hogan express the point perfectly when they put us in the position of the man in the street, whose reaction to a story of some harm to Society is to say *There ought to be a law against it.*[1]

This is not a textbook for lawyers but for aviation professionals. Nevertheless, Smith and Hogan's observation make a valuable point that lawmakers have taken a convenient path of relying disproportionately on criminal law by making strict

DOI: 10.1201/9781003489696-1

liability offences in the form of regulations, which dispense with the capricious process of proving all the elements of a crime under general law.[2]

So a crime is whatever Society says it is. People elect their lawmakers on the promises that are made to them, and it is this which underpins the concept of Justice, upon which laws are made and Prosecutions conducted. It is the evolutionary process of criminal accountability, defined and applied by rules that differ according to the normative ethics of changing societies, which has forged the process which the global aviation community identifies as the phenomenon of criminalisation. It has developed not by the intent, but by the harm caused, and certainly should only be applied when it is fair to do so.[3]

What is harm? Harm must always be judged in light of, and, therefore, must follow the determination of, culpability, or blameworthiness on the part of the offender. Such harm would include harm to the individual, who may suffer death, personal injury, psychological stress or financial loss; in other words, issues which are personal to the victim in question, and cause a particular impact which the Court should reasonably take into account when determining the seriousness of the offence. The harm may be caused by the following, as mere examples:

- Defects in manufacturing
- Improper maintenance of the aircraft
- Equipment failure
- Pilot error, negligence or recklessness
- Violation of safety standards set forth by the region's CAA
- Negligence of employees or air traffic controllers

What is fair? Fairness demands equal rights for all individuals, as a principle of justice which must be satisfied before other political interests are satisfied. But the controversy surrounding the criminalisation of professionals in the aviation industry is founded on the perception that they are burdened with criminal accountability beyond that accorded to any other private citizen, by virtue of their risk management functions, and the fatal consequences if they get it wrong. In this chapter, we shall examine how criminal liability works, giving us a picture of the basic characteristics of a crime.

Evolving trends in legislation add another layer to the question of fairness, in statutory wrongs that dispense with the need for proof of a guilty mind, and focus on the State's role of punishing the offender without the need for proving all the elements of a crime. Rationalising those trends in legislation can be complicated, though; in the UK's Civil Aviation Act 2012, the CAA has been given powers to make use of alternative civil sanctions alongside existing criminal penalties, blurring the distinction between criminal and civil law. There is no doubt that the main purpose of the provision is to save government money on Prosecutions, but it does offer an option to criminal sanctions, which will be discussed in Chapter 7.

THE ESSENCE OF CRIMINAL ACCOUNTABILITY

Society changes constantly, evolving in response to internal change and external stimuli and, with it, the concept of a risk which might threaten the security of that

Society has evolved. As a result, the late twentieth-century concept of a risk Society has been well established, in which Society has become more concerned with public safety than with economic inequality.[4] The aviation professional will immediately grasp the relevance of this to their own position.

The criminal accountability of a Defendant is defined and regulated by the State's domestic law, whose sovereignty is guarded jealously in accordance with Article 2(1) of the United Nations Charter: *The Organization is based on the principle of the sovereign equality of all its Members*. In other words, a State's sovereignty is defined by its power to make its own laws and no other nation can interfere in its own administration of justice. That being said, a State's criminal law, whether set out in a Statute or a Code, or indeed by the Common Law, is not necessarily set in stone, because Society is not set in stone. In other words, in a democracy what the People want the People get, so if attitudes change over years, the approach taken by the law will change too.

How the Defendant must answer to such laws will depend on what is written in the State's system and management procedures. In the English system, the Prosecution must establish both elements of the crime in question:

1. The *actus reus* contains all the elements in the definition of the crime except the Defendant's mental element. It is generally, but not invariably, made up of the Defendant's conduct and sometimes the consequences of that conduct, as well as the circumstances in which the conduct took place.
2. The Prosecution must also establish the Defendant's Guilty Mind, or *mens rea*. The primary function of the Prosecution case therefore must be to establish the Defendant's intention to commit the crime.[5] A result is intended when it is established beyond reasonable doubt that the Defendant deliberately and unjustifiably pursues a course of action while consciously disregarding any risks flowing from such action.

Like the Ten Commandments, these elements are few but have stood the test of time, and the noticeable absence of a political slant or, indeed, of bias, clearly establishes the founding principle that puts justice to the individual above political governance of Society as a whole. The phenomenon of criminalisation only began to evolve after the exploitation of these elements were relaxed to meet the political demands in the administration of justice for a more certain conviction in particular cases, which puts the characteristics of a crime in the context of the aviation professional[6] into sharp focus regarding the *mens rea*. In a criminal Prosecution, the mental element is required to establish proof of guilt beyond reasonable doubt; failure to establish a guilty mind according to the crime defined in law will lead to an acquittal. The exception, however, so often giving rise to criminalisation, arises in the case of strict liability offences, which were originally promulgated in the nineteenth century to provide health and safety regulations in the workplace. Once challenged with such an offence, it is for the Defendant to prove on the balance of probabilities – that is, more likely than not – that they discharged their duties with due diligence; if the case is proved against them, then this will give rise to liability to pay a penalty, without the trappings of a criminal offence to which blame was attached, such as a case involving criminal negligence.

This was qualified by Section 8 Criminal Justice Act 1967, which provides that, when assessing whether the Defendant had the necessary *mens rea*, the Jury shall not be bound to infer that the Defendant intended or foresaw the result of his actions merely because it is the natural and probable result of those actions, but by the subjective test of whether he actually did intend or foresee that such result would ensue, but the Jury must apply the test objectively by *drawing such inferences from the evidence as appears proper in the circumstances* (the Statute's own words). Whether the Jury understands what they have to do is an entirely different matter.

This reference to the 1967 Act illustrates the point that it is up to Parliament as to how it qualifies the requirement of *mens rea* in general; how particular issues are refined, may be established by decided cases. A notable example of this in context is recklessness, qualifying the requirement of intent, which will incriminate a person in a particular offence depending on his awareness of a risk given the circumstances or the result that may be apprehended from the circumstances.[7]

All that being said, the Prosecution still must prove the case against the Defendant beyond reasonable doubt, and this can only be achieved by the submission of evidence, which even after rigorous scrutiny by the Defence to put such evidence in doubt leads the Jury to be sure of their verdict.

CRIMINAL PROSECUTIONS

The Investigation

We may start by looking at the UK perspective, which has been followed in many systems around the world; then we can draw comparisons and distinctions with other systems.

In the UK, the Police and Criminal Evidence Act 1984 (PACE) implemented the recommendations set out by the Royal Commission on Criminal Procedure. Its purpose is to unify police powers under one code of practice and to balance the powers of police with the human rights of individuals. It provides for codes of practice which police officers should consider and observe when carrying out various procedures associated with their work. The act addresses issues such as arrest and detention, and, in particular, legislates on the admissibility of evidence that the police have gathered.

In the case of an aviation casualty, which demands experts in their technical field to gather evidence, the Air Accident Investigation Branch (AAIB) examines and investigates all types of civil aviation accidents and serious incidents involving UK-registered aircraft worldwide, and aircraft in UK territory, but this is an entirely different process to that involved in a criminal investigation to which PACE refers.

In English law, the Statutory Instrument extending the authority to the Executive to investigate accidents and serious incidents currently is the Civil Aviation (Investigation of Air Accidents and Incidents) Regulations 2018, issued under powers which Parliament delegated to the Secretary of State under the Civil Aviation Act 1982, and requiring the Chief Inspector of Air Accidents to cause a safety investigation to be conducted, insofar as it is practicable to do so, where an accident or serious incident occurs to a UK aircraft in any State that is party to the Chicago Convention.

Its legislative context is directly founded on the European-wide, appropriately named Investigation and Prevention of Accidents and Incidents in Civil Aviation, under European Regulation 996/2010,[8] which defines an *accident* as an occurrence associated with the operation of an aircraft which:

- takes place between the time any person boards the aircraft with the intention of flight until such time as all such persons have disembarked, in which a person is fatally or seriously injured; or
- sustains damage or structural failure which adversely affects the structural strength, performance or flight characteristics of the aircraft, and would normally require major repair or replacement of the affected component; or
- is missing or is completely inaccessible.

A *serious incident* is defined as one involving circumstances indicating that there was a high probability of an accident and is associated with the operation of an aircraft.

The sole objective of AAIB accident investigations is to determine the circumstances and causes of the accident with a view to preserve life and avoid accidents in the future, not to apportion blame or liability; hence it has an entirely different character, and a fundamentally different function, to the criminal investigation.

As PACE defines police powers to gather evidence, so the function of the AAIB is defined to gather the evidence in aviation accidents and serious incidents, which will be collected in a report and then all evidence may be considered by the CAA, which is tasked by the Department for Transport to investigate and prosecute breaches of aviation safety rules and some aviation related consumer protection and health and safety requirements. The CAA prosecutes both companies and individuals for breaches of the legislation it is tasked with enforcing. Investigations into alleged breaches of the law are carried out by the Investigations & Enforcement Team, in accordance with its Code of Practice.

Although defined as a Crown Prosecutor in its own right, the CAA may refer serious cases, demanding greater expertise in prosecutions of serious crimes, to the Crown Prosecution Service (CPS). Examples of offences include violations of the following:

- The Aviation Security Act 1982;
- The Civil Aviation Act 1982;
- The Aviation and Maritime Security Act 1990;
- The Railways and Transport Safety Act 2003; and
- The Air Navigation Order 2016.

The Crown Prosecutor therefore prosecutes criminal cases that have been investigated by the police or the AAIB along with other investigative organisations in England and Wales. The Crown Prosecutor is independent, and makes its decisions independently of the police and government. It states its duty to ensure that the right person is prosecuted for the right offence, and to bring offenders to justice wherever possible.

This summarises the investigation of the process for criminal purposes in the UK, which stands separate and apart from the technical investigation of air accidents, established under the Chicago Convention on International Civil Aviation which states that *the sole purpose of protecting safety information from inappropriate use is to ensure its continued availability so that proper and timely preventive actions can be taken and aviation safety improved.*[9] It is this from which the AAIB takes its lead. The Convention is careful to preserve the sovereignty of States in the unfettered administration of justice and their own criminal Prosecutions systems. The most it can require is that State laws *should ensure that a balance is struck between the need for the protection of safety information in order to improve aviation safety, and the need for the proper administration of justice.*[10]

The European Union underlined key issues which currently lie in its current Regulation 996/2010, noted above, but fails to address a problem which has manifested in cases in some States in which the judicial authorities play a more significant role in accident investigation, with a view to identifying criminal accountability, and the accident report prepared under Annex 13 has been used in evidence in criminal Prosecutions. The EU Regulations provides[11] that *when a judicial investigation is also instituted, the investigator-in-charge shall be notified thereof.* In those circumstances, *the relevant information collected in the safety investigation shall be shared with those authorities immediately and upon request, relevant material may also be transferred to those authorities.*

While efforts may be made to safeguard the independence of the investigators, the obvious risk arises that unlawful interference will follow, and the Regulation merely obliges the responsible government minister to cooperate with each other through advance arrangements, and *ensure respect for the independence of the safety investigation authority and allow the technical investigation to be conducted diligently and efficiently.* The dilemma presented to the professional complying with their obligation to co-operate with the accident investigation may conflict with their right to protect themselves from self-incrimination, which is a well-founded right under English Common Law but is absent from the European Convention on Human Rights. We shall discuss this in more detail later in the book.

The Decision to Prosecute

At the head of the CPS is the director of public prosecutions (DPP), who has the power under Section 10 Prosecution of Offences Act 1985 to issue a Code for Crown Prosecutors giving guidance on general principles to be applied by them

a. in determining, in any case
   i. whether proceedings for an offence should be instituted or, where proceedings have been instituted, whether they should be discontinued; or
   ii. what charges should be preferred; and
b. in considering, in any case, representations to be made by them to any magistrates' court about the mode of Trial suitable for that case.

The Crown Prosecutor therefore is responsible for deciding on whether to proceed with criminal proceedings against a potential Defendant, by which it follows tests contained in guidance.

Evidential Test

The decision to commence legal proceedings against a person is first determined by considering whether there is sufficient evidence to provide a realistic prospect of a criminal conviction. Such decisions are based on the:

- relevance of available evidence
- admissibility of that evidence
- assessed reliability and availability of Prosecution witnesses

Public Interest

If a matter passes the evidential test, then the Crown Prosecutor will consider if a Prosecution would be in the public interest. There is no definitive guidance as to when it may or may not be in the public interest to prosecute. Each decision will be on the individual facts of the case, but the factors contained in the following sub-paragraphs will be considered when coming to a decision.

Factors to Be Considered

There are a number of factors to be considered, such as:

- a conviction is likely to result in a significant sentence
- there is a history of recurring conduct by the person
- the person is or was in a position of authority or trust
- the alleged offence was of a serious nature

Serious Nature

To determine whether or not the offence is of a serious nature, the Crown Prosecutor must consider further matters, such as:

- the course of conduct
- the duration of the unlawful conduct
- the impact of the conduct on the alleged victim or others
- the personal harm or suffering of the alleged victim or others
- reckless disregard for the interests of the alleged victim or others

Other Factors

Where the evidential test and public interest assessment has been satisfied, a criminal Prosecution will be unlikely to be pursued if any of a number of factors are present, such as:

- the circumstances were beyond the person's control
- other relevant matters to be taken into consideration such as a lack of witness co-operation
- the commission of the alleged offence was the result of a genuine mistake or misunderstanding
- to bring proceedings would be an abuse of process

The glaring conclusion from this is the discretion which the Prosecutor exercises in their decision to bring criminal proceedings, which goes to the heart of the controversy as to who draws the line in criminalisation.

The Trial Process

In the English legal system, serious criminal cases will be heard in the Crown Court before a Jury; many jurisdictions such as the United States, Australia and India broadly following the UK's process. In fairness, while Juries may be somewhat capricious in their decisions, the reason for their verdict will not be questioned, save in very exceptional circumstances. Many States, however, adopt a system derived from Roman law, with its source in the character of Roman republican government, defining the nature and development of Roman law, and its role in international law today. Roman Civil Law is practised throughout the countries of South America and Continental Western Europe; confusingly, some States have a hybrid system, such as Mauritius, whose constitution defines its criminal process under the English legal system but, in practice, is much influenced by French law.

That being said, there is close similarity in the general principles between the two systems in criminal law, largely thanks to international treaties, particularly in civil aviation; the key difference is the Trial process. In the English process, it is adversarial, in which the Prosecution and Defence are represented by an advocate, whose function is to see the dispute resolved in terms of their party's best interests. In the Trial of a matter, the advocates present their cases in law, and support their legal argument with the facts, in the form of evidence. The Judge manages the process on the foundation of the Rules of Natural Justice, now enshrined in the European Convention on Human Rights, according to the law which is argued before them (and often with them, in order to achieve complete understanding of the law to be applied). In criminal trials in the Crown Court, the evidence is delivered to the Jury – consisting of 12 peers without necessarily any legal training, who hear the evidence in chief of the witnesses, and their responses to cross-examination. Submissions in law are made to the Judge in the Jury's absence, for a consideration of the law is not within their remit. All they need to know is the substantive law which makes a persuasive case on the evidence which they hear.

The fundamental difference between the English and French systems is that the French system reflects a Continental trend in the administration of justice, the focus of which presents a system which is not adversarial but inquisitorial.

In the inquisitorial process, the parties must provide all relevant evidence to the Court, and then the Judges, not the advocates, call and examine the witnesses, although the advocates may suggest questions to the judges. A party who wishes to ask a question must submit a brief in the form of Articles of Proof, describing the matters in question, which then is lodged with the Court and the other side. The Court therefore acts as an investigative body – hence the term Inquisitorial.

Unlike English judges who are career lawyers and independent of the administrative system, Continental systems are staffed by professional bureaucrats who typically employ lower-ranking judges to gather and prepare the evidence in written form and present it to judges of higher rank. Written additions to the file continue as the case unfolds. The parties do have the opportunity to enter new evidence and argument and to respond to written pleadings filed by the other side. The basic stages in a Continental case can best be defined as follows:

- At a brief preliminary stage, pleadings are submitted and a Hearing (or Instructing) Judge appointed;
- An evidence-taking stage follows in which the Instructing Judge takes evidence and prepares a written summary;
- The decision-making stage follows, in which the Judges who will decide the case consider the record transmitted to them by the Instructing Judge, receive Counsel's briefs, hear their arguments and give their decisions.

In English criminal cases, the Prosecution must establish that the Defendant's *mens rea* coincides with the *actus reus* required for the crime. If ever there is an illustrative authority which blends instruction with entertainment, it is the non-marine case of *Deller*.[12] The Defendant persuaded a third party to purchase his car by representing that it was free from encumbrances. Deller, however, had previously borrowed money from a finance company which had taken a mortgage over the car as security and, so, he believed that he was telling a lie – satisfying the mental element of dishonesty. When it transpired that the car was mortgaged and, therefore, very unfree from encumbrances, he was charged with obtaining property by false pretences by intentionally misrepresenting a past or existing fact, contrary to Section 32 of the Larceny Act 1916. But it then transpired that the bill of sale for the transaction had not been registered in accordance with legal requirements and, therefore, the mortgage was void. The car, therefore, had been free from encumbrances all along, so the *actus reus* for the crime was not present, and Mr Deller's conviction was quashed on appeal.[13]

The requirement of *mens rea* remains the key issue in determining criminal accountability; this most ancient of bastions in criminal law was upheld by Lord Reid in *Sweet v Parsley*,[14] in the presumption that Parliament does not intend to make criminals of people who are not blameworthy for what they did,[15] quoting the venerable Brett J in *R v Prince*:[16]

> Upon all the cases I think it is proved that there can be no conviction for crime in England in the absence of a criminal mind or mens rea.

The demands made by the requirement of *mens rea* on criminal liability require some qualification of the general principle, if a person, who does not intend to cause a harmful result, takes an unjustifiable risk of causing it. Such is the foundation of recklessness, in which either the Defendant was aware of its existence or, in the case of an obvious risk, the Defendant failed to give any thought to the possibility of its existence;[17] or, having assessed the risk, nevertheless went on to take it.

Case Study: R v Henderson

In this Prosecution,[18] David Henderson, 67, of Hotham, East Riding of Yorkshire, was found guilty in November 2021 of recklessly endangering the safety of an aircraft, contrary to Articles 240 and 265 and Part 4 of Schedule 13 of the Air Navigation Order 2016. In addition, shortly before the Trial, he pleaded guilty to the offence of attempting to discharge a passenger, contrary to Section 1(1) Criminal Attempts Act 1981 and Articles 250 and 265 and Part 3 of Schedule 13 of the Air Navigation Order 2016.[19] The Judge emphasised that there was no evidence that the matters giving rise to the offences for which he had been found or pleaded guilty were responsible for the subsequent crash of the aircraft. Accordingly, the sentence which he imposed did not in any way take into account the fact that the aircraft crashed on the return flight, nor the tragic deaths of his passengers Mr Ibbotson and Mr Sala.

Henderson had organised the flights without having obtained the Air Operator Certificate which was required in law, and did not have in place any of the extensive safety or compliance systems required before the CAA would have considered granting an Air Operator Certificate, as a result of which the air safety regulations were fundamentally breached. Mr Justice Foxton held,

> I am satisfied that, in this case, you intentionally breached or flagrantly disregarded the applicable CAA regulations, and that you did so on a pre-meditated basis and for reasons of profit. With the exception of the expiry of Mr Ibbotson's type rating, I am also sure that you were reckless, rather than merely negligent, as to the risk created to the safety of the aircraft by your disregard of the applicable regulations, as manifested by your interactions with Mr Ibbotson in relation to the January flights which I have already set out and as must have been obvious to you given your very great experience as a pilot.
>
> So far as harm is concerned, what is relevant in the present context is the seriousness of the harm risked and the degree of risk. In many cases, including this one, endangering the safety of an aircraft will involve an unacceptable risk of death or serious Injury to anyone travelling on the aircraft, albeit I accept Mr Spence's submission that that outcome is far from inevitable. I accept, however, that in this case, the risk was a relatively low one, albeit one which could not safely be ignored.... I also accept that, on the evidence, you acted at all times to ensure that the aircraft itself complied with the regulations concerned with its fitness to fly, although I should record that the aircraft would have been subject to a much more onerous safety and compliance regime had you sought the Air Operator Certificate which was required for the business of carrying passengers for reward.

Henderson was sentenced to a period of 18 months' imprisonment, with an order to be released from custody no later than half of the way through the sentence and the remainder of the sentence will be served on licence, in the community.

What Is Evidence for Legal Purposes?

Central to the Law of Evidence is the concept of relevance. Evidence will not be admitted by the Court unless it goes some way to show, either directly or indirectly, that a fact in dispute did or did not exist. The Court will allow in evidence everything that is relevant, provided that it is admissible. Relevance and admissibility are separate concepts, and the Law of Evidence states that certain kinds of relevant evidence cannot be taken into account and, thus, will be excluded from the Trial. It is therefore apparent that, while the Inquiry might well hear evidence that may be dramatic or damning in the eyes of the public, it might be excluded from the Trial process.

The best-known exclusionary rule is the hearsay rule. The object of this rule is to limit a witness's evidence when offered to prove the facts asserted to things which the witness perceived with his own five senses: that is to say, things which they had seen, heard, felt, smelt and tasted. As evidence in an Inquiry such as a statutory accident investigation conducted by a Judge, hearsay would be allowed, because it is not a Court of Trial, but such evidence would not be permitted in criminal proceedings, with the consequence that the Prosecution could not refer it to the Jury at all.

Occasionally, the Law of Evidence allows a fact to be proved without evidence, such as when the Court takes judicial notice of a fact. Sometimes no evidence is required because a party makes a formal admission, but there is no doubt that such evidence would not have been allowed in criminal proceedings if that is intended to incriminate the Defendant in blame.

When evidence takes the form of a document, there are a number of rules which stipulate whether the document is acceptable quite apart from any issue of relevance or infringement of an exclusionary rule. This would certainly have been an issue to be argued in a criminal Trial arising out of the case of the *Herald of Free Enterprise* disaster, had there been one.[20] In fact, a criminal Trial was commenced, indicting P & O European Ferries on a charge of corporate manslaughter, but after a half-time submission, the Judge ruled that the Prosecution was not in a position to satisfy a key element of the case against them and thus there was insufficient evidence upon which the Jury could safely convict. The case was duly dismissed.[21]

The Laws of Evidence are complex, and a Jury may struggle to rationalise the evidence with the law, as a consequence of which in many circumstances it would be unsafe to leave decisions involving weak evidence to the Jury. As a result, at the close of the Prosecution case, the Judge will allow a submission by the Defence of no case to answer when there is no – or, at least, insufficient – evidence upon which, if the evidence adduced were accepted, a reasonable Jury, properly directed, could convict. In such a case, the Judge must duly direct the Jury to return a verdict of Not Guilty. Of course, it would be for the Judge to determine whether there is evidence capable of giving rise to a duty of care, when they must give the Jury appropriate directions.

The Verdict

In English criminal law, the Jury must be satisfied beyond reasonable doubt that the Prosecution has met the demands which the act defining the offence places on the evidence, if they are to return a verdict of Guilty. If the Defence places reasonable doubt in their minds, then they must return a verdict of Not Guilty. Once the Jury has duly returned a verdict, its function has been completed and the Judge will be solely responsible for the sentencing of a Guilty Defendant. Sentencing Guidelines provide guidance on factors which the Court should take into account that may affect the sentence, setting out different levels of sentence based on the two issues of culpability and harm. This guidance will be discussed later in this chapter.

THE CONSTITUENT ELEMENT OF NEGLIGENCE

No such demands of criminal principles are required in a civil claim, however: merely, that the Claimant has suffered loss as a consequence of the Defendant's behaviour. The gulf between civil liability and criminal liability rests, ultimately, therefore, on how the law determines blameworthiness. If the differences between the two are irreconcilable, then the gulf must be unbridgeable, with the result that criminal accountability becomes misconceived. Having analysed the elements of criminal accountability, we therefore must consider civil liability in negligence in order to reconcile the two.

The Case of Donoghue v Stevenson[22]

In this defining case, the Appellant sought to recover damages from the Respondent, who was a manufacturer of aerated waters, for injuries she suffered as a result of consuming part of the contents of a bottle of ginger beer which had been manufactured by the Respondent, and which contained the decomposed remains of a snail. The ginger beer had been purchased for the Appellant by a friend in a café and arrived at the table in its customary bottle of dark opaque glass, when the Appellant had no reason to suspect that it contained anything but pure ginger beer; having had some of the contents poured into a tumbler, which she drank quite uneventfully, her friend was then proceeding to pour the remainder of the ginger beer into the tumbler when a decomposed snail accompanied the rest of the contents out of the bottle. The Appellant claimed damages for alleged shock and severe gastro-enteritis. The Appellant further averred that the ginger beer was manufactured by the Respondent to be sold as a drink to the public (including the Appellant); that it was bottled by the Respondent and labelled by him with a label bearing his name; and that the bottles were thereafter sealed with a metal cap by the Respondent. She further averred that it was the duty of the Respondent to provide a system of working his business which would not allow snails to get into his ginger beer bottles, and that it was also his duty to provide an efficient system of inspection of the bottles before the ginger beer was filled into them, and that he had failed in both these duties and had so caused the event.

It was Lord Atkin's speech which defined the modern principle, introducing the concept of the *neighbour* as the party to whom the Defendant owed a duty of care:

> You must take reasonable care to avoid acts or omissions which you can reasonably foresee would be likely to injure your neighbour. Who, then, in law is my neighbour? The answer seems to be – persons who are so closely and directly affected by my act that I ought reasonably to have them in contemplation as being so affected when I am directing my mind to the acts or omissions which are called in question.[23]

Lord Atkin's colleagues illuminated his observation with their own analysis of this redefined principle of duty. *Donoghue v Stevenson* brought evolution, not revolution, to the law of negligence; nevertheless, it was this humble bottle of ginger beer which provided the hard case that led to the illumination of the principles underpinning the tort of negligence, although the principles themselves had not changed. We can summarise this with a rule, containing three basic principles of liability in civil law, which the Claimant must establish on the balance of probabilities:

1. that the Defendant, the party alleged to be negligent had a duty to the injured party to avoid acts or omissions which might cause them loss or damage – for example, the Pilot in command, who is responsible for safety in the operation of the aircraft, clearly owes a duty of care as a result as regards passengers or other lawful visitors to the aircraft, whether or not they are contracting parties or are conducting their duties in the context of airport operations.
2. that the Defendant was in breach of that duty of care, generally by showing that their conduct fell below the reasonably expected standard to be owed by such persons in those circumstances.
3. the Claimant must have suffered damage as a reasonably foreseeable consequence of the Defendant's breach of duty. It is not necessary for the breach of duty to have been the sole or even the main cause of death, provided that it contributed significantly to the victim's death.

Damages are awarded in compensation with the purpose of putting the Claimant in the position he would have been in, had the negligence not taken place – so long as that is all that the measure of damages does; they certainly cannot put the Claimant in a better position than he would have been.[24]

This gallop through the general principles of negligence starkly demonstrates the foundation principle characterising the tort that the Defendant's state of mind – their *mens rea*, essentially – is not the determining factor in their culpability.

Courts would still struggle with the test to establish the duty of care in the first place, though, in the absence of a statutory obligation or contract. It was not until 1990[25] that the House of Lords set out its three-fold test for defining the duty, upon which the downstream consequences of the case would lie:

- Harm must be reasonably foreseeable as a result of the defendant's conduct (as established in *Donoghue v Stevenson*);
- The parties must be in a relationship of proximity; and
- It must be fair, just and reasonable to impose the duty.

The duty of care must have existed at the time of the loss, but that need not necessarily be immediately recognisable; the question to be addressed is whether it was foreseeable as a result of what the Defendant did or did not do. Civil aviation professionals may question themselves to determine whether they might be held accountable in such circumstances. When a person has created or contributed to the creation of a state of affairs which they know, or ought reasonably to know, has become life threatening, the duty arises on them to act by taking reasonable steps to save the other's life will normally arise. In the case of *R v Evans*,[26] the Lord Chief Justice stated obiter:[27]

> The duty necessary to found gross negligence manslaughter is plainly not confined to cases of a familial or professional relationship between the Defendant and the deceased. In our judgment, consistently with Adomako and the link

> between civil and criminal liability for negligence, for the purposes of gross negligence manslaughter, when a person has created or contributed to the creation of a state of affairs which he knows, or ought reasonably to know, has become life threatening, a consequent duty on him to act by taking reasonable steps to save the other's life will normally arise.

The awful truth revealed by Michaelides-Mateou and Mateou[28] is that aviation professionals have little knowledge and understanding of the law of negligence and the implications raised by evidence of negligence when it finds itself as a material factor in a criminal Trial. Ignorance of the law is no defence, of course, but ignorance is also the midwife of fear, leading to the anticipated need to conceal errors in the investigation process, consequently compromising and, very possibly, defeating the Just Culture policy in order to avoid criminal accountability.

VICARIOUS LIABILITY

In civil aviation, there inevitably stands a risk that the negligence of an employee might incriminate their employer. Vicarious liability holds a person or company responsible for acts or omissions committed by others or by their employees. It is particularly valuable in the recovery of damages, for the employer will likely be in a better position to pay, whether or not through a policy of insurance. In this respect, the Pilot in command has a special contractual relationship with the owner at common law, arising out of their common interest in the success of the flight, which could be relied upon to maintain the bond between them. But the changes in the pattern of airline risk management in recent years have responded to the consequences of litigation that encourage the corporate owner to distance themselves from the employee, as Mark Forkner discovered in the Boeing case.[29] The downstream consequence is that the corporate owner very well have a vested interest in avoiding legal accountability by seeking to defend itself from vicarious liability. Such a defence includes establishing on a balance of probabilities that the employee's act was so far outside the idea of what was expected in the course of their employment that the employer could not be liable, or to show that the employer took all reasonable steps and precautions to avoid the commission of the offence or unlawful act by the person concerned.

Since the case of *Lister v Hesley Hall*,[30] the application of vicarious liability can be drawn very widely, to include unauthorised, intentional acts or omissions committed by an employee acting for their own benefit, as far as there exists a connection between the acts or omissions and the work for which they were employed to undertake in the course of their employment.

Given these elements of civil liability, we can now analyse just what is meant by an offence by which negligence becomes a crime.

WHAT IS CRIMINAL NEGLIGENCE?

In the UK, there is no crime, per se, of criminal negligence, but the modern tort of negligence has had a critical, and potentially critically damaging, effect on the Defendant's criminal accountability, most notably in the matter of manslaughter. Not that such a thing can be explained easily; in his paper "The Homicide Ladder" in the *Modern Law Review*,[31] Victor Tadros made psychology the interrogator of the question of human behaviour when it comes to putting manslaughter into the context of normative ethics. The conclusion finds that, when a sudden death takes place, it is human nature for people to find somebody to blame, and the reasoning behind the conclusions drawn may not bear any similarity to principles of law. As a result, the blame process may not be mapped onto any legal argument defining the characteristics of the crime of manslaughter, which means that the normative ethics of Society present a hazard to the process of a fair Trial by applying practical steps that do not necessarily satisfy the demands in theory of justice. If Society demands redress, therefore, the law which is put in place to serve that demand must stand up to scrutiny according to basic concepts of fairness in criminal accountability. Manslaughter has proved particularly vulnerable to such scrutiny; indeed, the very starting point has proved troublesome, for a definition has been notoriously difficult to establish. In *Andrews v DPP*,[32] Lord Atkin went further and held out manslaughter as being the most difficult of all crimes to define. While the guilty mind in murder requires an intention to kill (malice aforethought), even if that intention is benign, manslaughter essentially turns on an absence of an intention to kill. The really difficult part about manslaughter, however, is establishing the factor which must still be proved, that of unlawfulness.

A succession of cases over decades struggled with the need to define just how evidence of the Defendant's state of mind should underpin gross negligence, until, in 1995, the case of *R v Adomako*[33] established that the Defendant can be convicted of gross negligence manslaughter in the absence of evidence to their state of mind. In this case the Defendant was the anaesthetist in an eye operation on a patient. During the operation, the tube from the ventilator supplying oxygen to the patient became disconnected. The Defendant failed to notice the disconnection for some six minutes before the patient suffered a cardiac arrest, from which he subsequently died. The Defendant was charged with manslaughter. At his Trial, it was conceded on behalf of the Defendant that he had been negligent in the tortious understanding of the word and medical evidence was called by the Crown that the Defendant had shown a gross dereliction of care. The Judge directed the Jury that the test to be applied was whether the Defendant had been guilty of gross negligence. The Defendant was convicted.

Hearing the Appeal, the Lord Chancellor, Lord MacKay, referred to the opinion of Lord Hewart CJ in *R v Bateman*[34] in underpinning his conclusion that the criminal law requires a fair and reasonable standard of care and competence in individuals in the position of Mr Adomako, according to the evidence required to establish liability in the civil tort of negligence. This, alone, creates a serious problem in the judicial process, for, as has been noted, in civil cases, the Claimant must persuade the Court that it has proved on the Balance of Probabilities the requisites for establishing liability, whereas in criminal cases, the Prosecution must persuade the Court that the case against the Defendant is proved beyond reasonable doubt, by applying the evidence to the body of criminal law with all those characteristics special to it, that is, including the *mens rea*.

As if the mischief in the Court's decision were not clear enough, however, the Judge summed up the Prosecution's task as taking the requirements for establishing the tort and applying an objective test to establish the crime. In words which were astonishingly unambiguous, the Judge directed the Jury that, to establish criminal liability on the charges in the Indictment, the Prosecution had to satisfy them that the matters that were simply necessary to establish liability in the civil tort of negligence had been proved. The only factor which went beyond this level was that the facts had to persuade the Jury that the negligence (or *incompetence* – the Judge's word) of the Accused *went beyond a mere matter of compensation and showed such disregard for the life and safety of others as to amount to a crime against the State and conduct deserving punishment.*[35]

The Judge's observations on the application of civil liability massively outweigh that devoted to criminal liability; but, to summarise the principle in this case, the Jury needs to consider whether:

1. the Defendant owed a duty of care to the deceased; and
2. he was in breach of that duty; and
3. the breach was so grossly negligent that it should be seen as criminal.[36]

This third factor introduces into the civil test the criminal issues contained in the passage italicised above. Lord MacKay defined gross negligence as depending

> on the seriousness of the breach of the duty committed by the Defendant in all the circumstances in which he was placed when it occurred and whether, having regard to the risk of death involved, the conduct of the Defendant was so bad in all the circumstances as to amount in the Jury's judgment[37] to a criminal act or omission.[38]

The test for causation in criminal cases was succinctly put by Lord Woolf MR in 1998:[39]

> In relation to… manslaughter it is an essential ingredient that the unlawful or negligent act must have caused the death at least in the manner described. If there is a situation where, on examination of the evidence, it cannot be said that the death in question was caused by an act which was unlawful or negligent as I have described, then a critical link in the chain of causation is not established. That being so, a verdict of unlawful killing would not be appropriate and should not be left to the Jury.

It is not the function of the Jury to evaluate competing causes or to choose which was dominant, provided they are satisfied that the Defendant's actions could fairly be said to have been a significant contribution to the victim's death.[40]

The whole subjective issue of the Defendant's state of mind, their *mens rea* for the crime, has therefore been put into the form of an objective test determined by the circumstances and grafted on to the evidential test required for liability in civil proceedings. Of course, the law was created to meet the demands of Society, not the other way around and, so, the Courts must apply the law according to Statutes and precedents. Without the benefit of a dedicated Statute, the Common Law created by precedents must be upheld. In this context, the elements of manslaughter by gross negligence were stated concisely by the President of the Queen's Bench Division in *R v Rudling.*[41]

> We can summarise the law shortly. The critical ingredients of gross negligence manslaughter can be taken from R v Prentice, Adomako and Holloway [1994] QB 302 in this court and Adomako [1995] 1 AC 171, [1994] 99 Crim App R 362 in the House of

> Lords as well as R v Misra [2005] 1 Cr App R 21. They can be summarised as being the breach of an existing duty of care which it is reasonably foreseeable gives rise to a serious and obvious risk of death and does, in fact, cause death in circumstances where, having regard to the risk of death, the conduct of the Defendant was so bad in all the circumstances as to amount to a criminal act or omission (see Adomako [2005] 1 Cr App Rep at 369). The elements of GNM were set out by the House of Lords in R v Adomako [1995] 1 AC 171.

In the case of *Honey Maria Rose v R*,[42] the Court confirmed the ruling in *Rudling* and held, obiter, that the question of whether there was a serious and obvious risk of death must exist and be assessed with respect to knowledge at the time of the breach of duty. But a recognisable risk of a serious consequence is not the same as a recognisable risk of death. An obvious risk is a present risk which is clear and unambiguous, not one which might become apparent on further investigation; such an issue must be considered during the risk assessment process. A brief case study illustrates this in the aviation context.

Case Study: Crash over Newton, Wisconsin 1997

In the afternoon of 15 March 1997, a Beechcraft Bonanza light aircraft took off and began a formation flight with a Basler BT-67 commercial airliner from their home aerodrome at Oshkosh, Wisconsin. The Basler is one of the industry's most reliable aircraft, combining the airframe of the Douglas DC-3 Dakota, with the power of new generation turboprops, and engineered overall to assure that every component, assembly or system is either new or the equivalent of new. A Dakota Pilot once said: *You can wreck a Dak, but you can't wear it out!*[43] and the same is true today of the Basler; a technical defect in an aircraft maintained and flown according to the manual is exceptionally unlikely (although not unknown, hence the Boeing 737 cases, as we shall see).

Both aircraft were owned by the same operator, and the flight had been planned in order to get flying time and aerial photographs. There was some light to moderate turbulence but no evidence reported of more inclement weather. At 15.28, they suffered a mid-air collision, supported by eye-witness evidence that the aircraft were some 500 to 700 feet above ground level, flying close together heading north. The witness informed the Safety Board investigation that he was standing in clear view talking to a friend, as he saw the Basler flying straight and level, with the Beechcraft just to the west of it, when it *then hit the big plane near the middle.* Another witness saw the light aircraft circling behind the Basler before it crossed over the top when it hit the top of the BT-67. About five seconds after impact, the DC-3 gently rolled or turned westbound (apparently descending and gaining airspeed); the left wing then came off, followed by the right wing about two seconds later.

The Safety Board's findings[44] established that the flight control system had been severed, which made it impossible to control the aircraft, and it crashed in an uncontrolled descent. The report concluded that the collision occurred as a result of failure of the Beechcraft Pilot to maintain clearance from the BT-would 67, while carrying out its photography operation.

There were no survivors, as is frequently the case in aviation casualties, so no proceedings followed to establish the extent of negligence; however, the facts presented by the eye-witness evidence led to the assessment of an obviously present risk which is clear and unambiguous, not one which might only become apparent after further investigation. The question now follows as to whether the facts demonstrated an error of judgment that would be serious enough to amount to gross negligence.

WHAT IS THE GROSS (OR CRIMINAL) NEGLIGENCE ELEMENT?

We have encountered this phrase repeatedly, yet it was not until the case of *Camarata v Credit Suisse*[45] in 2011 that the concept was adequately defined in English law, confirming the element of the parties' intention in interpreting a contract, rather than looking at the strict words, to divine its meaning. In this case, though, there was no element of criminal accountability, but a far simpler issue in which the parties had already separated the concept of 'gross' negligence from ordinary negligence in the agreement, so clearly, they had intended that there be a difference. By way of its reasoning, the Court observed that the adjective 'gross' embraced not only the conduct undertaken with actual appreciation of the risks involved but also serious disregard of or indifference to an obvious risk.

Various terms have been used to describe the type of conduct that may amount to a crime of gross negligence. The 2004 case of *Misra*[46] reached the Court of Appeal which approved the Trial Judge when he said:

> Mistakes, even very serious mistakes, and errors of judgment, even very serious errors of judgment are nowhere near enough for a crime as serious as manslaughter to be committed.... The defendant's conduct must fall so far below the standard to be expected of a reasonably competent and careful person that it was something truly, exceptionally bad.

In other words, the mental element of a guilty mind is not in issue. Lord Hailsham famously re-stated Brett J's maxim when he held obiter in *Haughton v Smith*[47] that *the deed does not make a man guilty unless his mind be also guilty.* This laconically describes the gulf between civil and criminal law, which has become blurred by cases such as *Adomako* and, more recently, by the Corporate Manslaughter and Corporate Homicide Act 2007, which, however, has had an extremely unsafe start in life.

A great deal of disagreement, not to say confusion, bedevils a rational analysis that seeks to bridge the gulf, partly because criminal pleading would argue that, had the risk not formed part of the Defendant's mental element, then there would not be the *mens rea* for a crime. It is this state of mind which presents the hurdle to a solution, for the determination of the Defendant's state of mind must necessarily be established by a subjective test, analysing guilt by the individual's intent, rather than by the test which places the Jury's focus on the consequences of the individual's acts, with an objective consideration of reasonable foreseeability which a civil court would have to consider in negligence, leading to a conclusion on whether it was *fair, just and reasonable*[48] to impose the duty of care that was allegedly broken.

The application of the principle in the context of negligence must, therefore, rest upon the definition of the appropriate test. Having discredited thoroughly the objective nature of the test in *R v Adomako*[49] and condemned it in *Williams v Natural Life*

Health Foods,[50] a subjective test must offer the only alternative. The only outstanding issue that can remotely fall to be determined is whether the Defendant's 'inner posture' had been 'reasonable.' The question of reasonableness was defined in *R v K*,[51] when the ever-redoubtable Lord Bingham put the position that the Defendant's belief must be established as honest and genuine; if so, it need not necessarily be reasonable by any objective standard, although, the more unreasonable the belief, the less likely it is to be accepted as genuine.[52]

In essence, therefore, criminal liability will depend upon the question as to whether the Defendant's state of mind amounted to recklessness. If the Pilot in command genuinely does not apprehend the risk that a crime could occur as a result of their action, then the question of a guilty mind does not arise in the general principles of criminal accountability, leaving them with civil accountability for the consequences of a breach of the duty of care in the tort of negligence; in what circumstances, therefore, could the facts in negligence give rise to a criminal state of mind justifying a Prosecution?

The test has been established under English law, in a succession of cases in criminal damage, culminating in the House of Lords decision in *R v G*.[53] While striving to interpret the Will of Parliament – which is the Court's prime function in the determination of criminal law – the House of Lords held a subjective standard to apply so that, a person who, without lawful excuse, destroys or damages any property belonging to another, either with intent or by recklessness, will be guilty if he is aware of the risk that a circumstance exists or will exist; or a result when he is aware of a risk that the result will inevitably follow; and it is, in the circumstances known to him, unreasonable to take the risk.

Such basic principles have evolved through a succession of judicial decisions, particularly as to recklessness. In fairness, these characterise the very essence of a hard case, which demands a consideration of the legal principles involved to ensure the best possible result, even if this means ignoring previously established rules. Naturally, the Judge will manage the process by drawing upon the arguments presented by the advocates – Cicero's sage observation is as relevant today as ever, that if cases could speak for themselves, no one would employ an advocate[54] – but, in hard cases, the Judge can influence the system.

The constant factor running through all of them, clearly, touches upon the Defendant's state of mind. In *R v G*,[55] Lord Bingham articulated the point beyond any doubt that conviction of a serious crime should depend on proof not simply that the Defendant caused (by act or omission) an injurious result to another but that his state of mind when so acting was blameworthy. Taking an obvious and significant risk by intention or recklessness would satisfy Lord Bingham of a guilty mind but not if the Defendant did not perceive the risk; whatever that made him, it did not make him a criminal.[56]

Pilot error with criminal negligence is a serious charge that can be brought against a Pilot in the event of an aviation accident. Some examples of this type of error include:

- Failing to properly check or maintain the aircraft before flight
- Ignoring weather or air traffic warnings and flying into dangerous conditions
- Operating the aircraft while under the influence of drugs or alcohol
- Failing to properly follow established flight procedures or regulations
- Neglecting to properly train or supervise other crew members

It is important to note that these are just a few examples, and that the specific circumstances of each case will be taken into account when determining whether criminal negligence was a factor. Additionally, it's worth noting that criminal negligence is a severe legal charge, and should not be taken lightly.

In *R v Sellu*,[57] the court quashed a conviction on two grounds. Cited with approval by Leveson LJ in the appeal of Honey Maria Rose,[58] he said:

> the Judge failed to direct the Jury sufficiently as to the line that separates serious or very serious mistakes or lapses from conduct which was truly exceptionally bad and was such a departure from that standard [of a reasonably competent doctor] that it consequently amounted to being criminal.

In *Sellu*, on 5 November 2013, the Defendant, David Sellu, a consultant specialising in colorectal medicine was convicted by a majority verdict of the manslaughter by gross negligence in 2010 of James Hughes, then aged 66 years (he was acquitted of perjury in relation to evidence given to the coroner about the death). He was sentenced to two and a half years' imprisonment. No appeal either against conviction or sentence followed but detailed further consideration of the case led new counsel (instructed by different solicitors) to pursue a wide-ranging appeal. Hearing the appeal, the obligation by the Trial Judge's summing up to the Jury was emphasised with great importance, to explain to the lay people in the Jury the seriousness of the departure from ordinary standards required by the concept of gross negligence. In this case, the Jury had not had the benefit of sufficiently detailed directions as to the concept of gross negligence contained within the offence of gross negligence manslaughter. That failure had been made worse by the way in which the expert witnesses had been allowed to trespass on matters of law and wrongly asserted gross negligence in their evidence, which was aggravated by the Judge's failure to lead the Jury on a route to reach a verdict which would have focussed their minds on the various stages to be considered. Accordingly, the conviction was unsafe and would be quashed, a rare example of a successful appeal against conviction.

THE EVOLUTION OF CRIMINAL ACCOUNTABILITY

It is an uncomfortable paradox that the modern civil tort of negligence has had a critical effect on criminal accountability, in the matter of manslaughter. Not long after the case of *Donoghue v Stevenson*, Lord Atkin heard the appeal of a Defendant who had driven a van above the speed limit and overtook another car; as he did so he struck a pedestrian and killed him.[59] Lord Atkin held out manslaughter as being the most difficult of all crimes to define. While the guilty mind in murder requires an intention to kill (malice aforethought), even if that intention is benign, manslaughter essentially turns on an absence of an intention to kill. The really difficult part about manslaughter, however, is establishing the factor which must still be proved, that of unlawfulness, and his conclusions may be applied to so many civil aviation cases today:

> There is an obvious difference in the law of manslaughter between doing an unlawful act and doing a lawful act with a degree of carelessness which the Legislature makes criminal.

Simple lack of care that would constitute civil liability in negligence is not enough. For purposes of the criminal law, there are degrees of negligence: and a very high degree of negligence is required to be proved before the crime is established.

A succession of cases over decades exposed the need to define just how evidence of the Defendant's state of mind should underpin gross negligence, until *Adomako* established that the Defendant can be convicted of gross negligence manslaughter without proving a guilty mind. The whole issue of the Defendant's state of mind, their *mens rea* for the crime, has therefore been put into the form of an objective test determined by the circumstances and grafted on to the evidential test required for liability in civil proceedings. In recent years, the rationale of criminal accountability has been simplified by the creation of offences of strict liability. This is a vital issue for us because breaches of most of the civil aviation rules are offences of strict liability.

Lord Atkin had defined the issue very clearly; but the law had to catch up with the demands of Society. The lawmakers devised a solution, in legislating for offences of strict liability, which produced a huge evolutionary step in criminal accountability, which are promulgated for the benefit of Society, overriding any disadvantage to the human rights of the person being charged in terms of whether their contravention was deliberate or accidental, therefore dispensing with the element of *mens rea*. The concept of strict liability is by no means new; it is founded on the procedure of making regulations for the benefit to the community which enforce certain kinds of behaviour or conduct. But such disregard for the human rights of justice to the individual demand very careful scrutiny.

The value of strict liability offences was identified as being so useful in civil aviation that it was embraced in Article 17 of the Montreal Convention:[60]

1. The carrier is liable for damage sustained in case of death or bodily Injury of a passenger upon condition only that the accident which caused the death or Injury took place on board the aircraft or in the course of any of the operations of embarking or disembarking.
2. The carrier is liable for damage sustained in case of destruction or loss of, or of damage to, checked baggage upon condition only that the event which caused the destruction, loss or damage took place on board the aircraft or during any period within which the checked baggage was in the charge of the carrier. However, the carrier is not liable if and to the extent that the damage resulted from the inherent defect, quality or vice of the baggage. In the case of unchecked baggage, including personal items, the carrier is liable if the damage resulted from its fault or that of its servants or agents.

It is noteworthy that the offence is one of strict liability, not absolute liability. Article 20 provides some element of a defence in this context:

> If the carrier proves that the damage was caused or contributed to by the negligence or other wrongful act or omission of the person claiming compensation, or the person from whom he or she derives his or her rights, the carrier shall be wholly or partly exonerated from its liability to the claimant to the extent that such negligence or wrongful act or omission caused or contributed to the damage. When by reason of death or

> Injury of a passenger compensation is claimed by a person other than the passenger, the carrier shall likewise be wholly or partly exonerated from its liability to the extent that it proves that the damage was caused or contributed to by the negligence or other wrongful act or omission of that passenger.

In addition to the provisions of the Montreal Convention, strict liability offences carry the general defence of due diligence. Even though the Prosecution may establish beyond reasonable doubt that the provisions of the offence have been violated, the Defence may prove on the balance of probabilities that they took all reasonable precautions and exercised all due diligence to avoid an offence being committed. The Australian CAA offers an example of a Pilot who committed an offence of low flying; if the Pilot could show that they believed that they were flying at a lawful altitude based on the readings on the aircraft altimeter, but it is discovered that the altimeter was faulty and they neither knew nor had any reason to know of the fault, that would be a defence.[61]

In 2010, the Law Commission published its Paper *Criminal Liability in Regulatory Contexts*,[62] supporting the proposition that the State's criminal justice system should not be exploited as the primary means of promoting regulatory objectives. It stated:

> 1. The criminal law should only be employed to deal with wrongdoers who deserve the stigma associated with criminal conviction because they have engaged in seriously reprehensible conduct. It should not be used as the primary means of promoting regulatory objectives.
> 2. Harm done or risked should be regarded as serious enough to warrant criminalisation only if,
>   a. in some circumstances (not just extreme circumstances), an individual could justifiably be sent to prison for a first offence, or
>   b. an unlimited fine is necessary to address the seriousness of the wrongdoing in issue, and its consequences.
> 3. Low-level criminal offences should be repealed in any instance where the introduction of a civil penalty (or equivalent measure) is likely to do as much to secure appropriate levels of punishment and deterrence.

Having endorsed the proposal for regulatory offences, the Law Commission proceeded to emphasise the importance of providing for a due diligence defence:

> The courts should be given a power to apply a due diligence defence to any statutory offence that does not require proof that the Defendant was at fault in engaging in the wrongful conduct. The burden of proof should be on the Defendant to establish the defence.... If this is accepted, then the defence of due diligence should take the form of showing that due diligence was exercised in all the circumstances to avoid the commission of the offence.

The Pilot in Command is the leader in the flight operation, whose responsibilities are defined in the Air Navigation Order 2016, reflecting the demands of European international law:

> 68. The Pilot in command of an aircraft is responsible:
>
> a. before every flight, for defining the roles and duties of each crew member;

b. for the operation and safety of the aircraft and for the initiation, continuation, termination or diversion of a flight in the interest of safety; and
c. during aircraft operations, for ensuring the safety of all crew members, passengers and cargo on board.

WHEN DOES PROFESSIONAL NEGLIGENCE BECOME A CRIME?

In American law, professional responsibility is defined within the concept of professional ethics for those who exercise a unique set of knowledge and skill as professionals.[63] Professional responsibility applies to those professionals making judgments, applying their unique skills, and reaching informed decisions for, or on behalf, of others, as professionals.[64] Professionals must be seen to exercise due care and responsibility in their areas of specialisation – their professions. What makes professionals unique is that the general public would not ordinarily be expected know in detail the skills and knowledge of a profession independently.[65]

In the civil aviation context, professional responsibility encompasses an array of the personal, corporate and humanitarian standards of behaviour, as expected of trained professionals and those who regulate and examine their conduct.[66] In fact, the foundation of this offence is that the degree of negligence needs to be very high before the conduct can be considered to be a crime. Hence, the Court must test the Defendant's conduct, the seriousness of the breach and the facts surrounding the circumstances before a verdict can be reached – and that will require the assistance of expert witnesses.

The circumstances in which the accident took place therefore are crucial in determining whether the professional conduct fell below the expectations reasonably to be expected of them; and expert findings may draw that into doubt.

THE INTERVENTION OF A JUST CULTURE

It is important to emphasise the context of the Just Culture in this chapter. The Chicago Convention on International Civil Aviation was signed on 7 December 1944 by 52 States and ratified three years later. Annex 13 addresses the protection of information from safety data collected in investigations and emphasises that *the sole purpose of protecting safety information from inappropriate use is to ensure its continued availability so that proper and timely preventive actions can be taken and aviation safety improved.* While the Convention is not designed to compromise the sovereign rights of a State in the administration of justice, it obliges States to ensure that *a balance is struck between the need for the protection of safety information in order to improve aviation safety, and the need for the proper administration of justice.*[67]

The imperative in the Chicago Convention demands that State Prosecutors should take steps to gather evidence to support their own investigations upon which to bring criminal proceedings, rather than attempt to use that obtained by the accident investigation team. The whole purpose of accident investigation is to prevent accidents, not to support the Prosecution of an offender. Eurocontrol, whose formal title is the European Organisation for the Safety of Air Navigation, has produced its model policy for criminal investigation and Prosecution of transport accidents *to provide directions regarding the criminal investigation and Prosecution of potential criminal offences resulting from aviation incidents or accidents that come to the attention of prosecutors so that both the safety investigation and judicial process can progress in parallel without either party acting in a manner prejudicial to the interests of the other.*[68] It particularly warned that the abuse of the system, in order to support criminal Prosecutions, would result in a loss of trust and cause harm to the process of improving air safety. In other words, a Just Culture policy should encourage all the participants to engage in the process of improving air safety, without fear of incriminating themselves in criminal liability.

REFORM OF THE SYSTEM IN THE UK

The Civil Aviation Act 2012 performs a number of functions. It contains:

- measures which modernise the regulatory framework for civil aviation in the United Kingdom through reforms to the legislative framework for the economic regulation of airports and the legislative framework of the CAA;
- measures conferring certain aviation security functions on the CAA;
- measures to reform the Secretary of State's powers to regulate the provision of flight accommodation.

More relevantly to this work, it also enables the CAA to make use of alternative civil sanctions alongside existing criminal penalties.

The CAA previously investigated and prosecuted aviation-related offences on behalf of the Crown, pursuant to arrangements to provide assistance to the Secretary of State under Section 16 Civil Aviation Act 1982. The act makes express provision for the CAA to institute criminal proceedings as part of its enforcement function under Section 20 Civil Aviation Act 1982 to ensure that costs associated with carrying out enforcement work, including Prosecutions, will be recoverable from the industry under the new charging scheme arrangements rather than from the taxpayer, as was the case under previous arrangements. The amended Section 20 provides supplementary provisions with respect to the functions of the CAA, namely that it shall *without prejudice to its powers apart from this paragraph have power to do anything which is calculated to facilitate, or is conducive or incidental to, the performance of any of its functions (including a function conferred on it by virtue of this paragraph).*

The power of the CAA to reach beyond the sovereignty of the UK is reflected in the following paragraph, that *it is hereby declared that the CAA has power to do elsewhere than in the United Kingdom such things as it considers appropriate for the purpose of performing any of its functions.*

This power conferred on the CAA particularly includes power to institute and carry on criminal proceedings in England and Wales or Northern Ireland, in place of the CPS – provided, of course, that it has the necessary talent and skills to do so, failing which it would instruct the CPS. Nevertheless, the CAA has prosecuting powers *in performing any function which the CAA is authorised by an Air Navigation Order to perform and which in pursuance of, or of an Annex to, or of an amendment for the time being in force of, the Chicago Convention falls to be performed on behalf of the Crown*; provided, that the powers do not exceed the obligations of the United Kingdom under an international agreement.

Section 99 of the Act enlarges on the Prosecution of offences allegedly committed by a corporate organisation *and is proved to have been committed with the consent or connivance of or to be attributable to any neglect on the part of any director, manager, secretary or other similar officer of the body corporate or any person who was purporting to act in any such capacity, he as well as the body corporate shall be guilty of that offence and be liable to be proceeded against and punished accordingly.*

This, conveniently, brings us to the subject of corporate accountability.

CORPORATE MANSLAUGHTER

Corporate accountability does not lend itself to precise definition, but may fairly be described in plain English as the legal obligation or duty to provide an account or justification for a corporate body's actions to whomever one is answerable. The 'whomever' has come to be known as the stakeholder; just who satisfies that definition naturally has been the subject of litigation in negligence for generations, but the Neighbour Principle in *Donoghue v Stevenson* must offer the best guidance in the context of this work: a stakeholder will be somebody so closely and directly affected by the Company's act that it should reasonably foresee that they would be so-affected.[69] With the implementation of the Corporate Manslaughter and Corporate Homicide Act 2007, corporate accountability for the crime of manslaughter has taken on statutory form, which satisfies the normative ethics of a Society that demands redress for catastrophic events such as the *Herald of Free Enterprise*, but at the cost of the nerves of corporate owners and operators.

THE EVOLUTION OF CORPORATE ACCOUNTABILITY

At common law, the test of a corporate body's criminal liability for manslaughter was dependent upon whether evidence can establish beyond reasonable doubt that a director of the company – part of the Controlling Mind and Will of the Company – was personally guilty of manslaughter. This identification doctrine was described in *HL Bolton v Graham & Sons Ltd.*[70]

> A company may in many ways be likened to a human body. It has a brain and nerve centre which controls what it does. It also has hands which hold the tools and act in accordance with directions from the centre. Some of the people in the company are mere servants and agents who are nothing more than the hand to do the work and cannot be said to represent the mind and will. Others are directors and managers who represent the directing mind and will of the company, and control what it does. The state of mind of these managers is the state of mind of the company and is treated by the law as such.

In *Tesco Supermarkets Ltd v Nattrass*,[71] three different judges gave three slightly different interpretations of who could be defined as participating in the controlling mind. Lord Reid stated that the following individuals were controlling minds of the company:

> The board of directors, the managing director and perhaps other senior officers of a company carry out the functions of management and speak and act as the company.

Viscount Dilhorne gave a narrower interpretation saying that a controlling mind is a person

> who is in actual control of the operations of a company or of part of them and who is not responsible to another person in the company for the manner in which he discharges his duties in the sense of being under his orders.

Lord Diplock stated that the people who form the controlling mind are those

> who by the memorandum and articles of association or as a result of action taken by the directors of the company in general meeting pursuant to the articles are entrusted with the exercise of the powers of the company.

Mr Justice Taylor addressed the issue in the *P & O Ferries* case, stating[72]

> when a corporation, through the controlling mind of one of its agents, does an act which fulfils the pre-prerequisites for the crime of manslaughter, it is properly indictable for the crime of manslaughter.

The undoubted catalyst in the development of Statute law lay in the case of the *Herald of Free Enterprise.*

The Case of the *Herald*

Summary of the Facts

By the summer of 1986, the board room of P & O Ferries was thinking hard about its long-term strategy, in order to keep its position in the market-place and satisfy the shareholders. Paradoxically, it was this decision which led to the Company taking over the Ownership and operation of the ill-fated *Herald of Free Enterprise*, just a month before her foundering. The *Herald* was one of three sisters, a modern roll-on/roll-off passenger/vehicle ferry grossing 7,950 tons, and one of the largest vessels of her type when she left her German builders in 1980. She had been designed at a time when her Owners, Townsend Thoresen, knew that every second and every penny counted in the cut-throat competition on the English Channel. Indeed, so thorough had her design concept been, that the linkspans at either end of her planned service, between Dover and Calais, were especially built to match the levels of the ship's car decks, in order to enable simultaneous loading of both decks and thus minimise the time required for turn-around in port.

The competition just carried on growing on the English Channel, one of the most crowded seaways in the world – and the Dover-Calais run was the most competitive crossing, because it was the shortest – just 22 miles long. While the UK's growing trade with the European Community undoubtedly contributed to the rising demand, competition was heating up thanks to a number of factors. In July 1984, The UK government announced the privatisation of the former State-owned ferry operator Sealink, which would lead to more pressure still on the competitiveness of the rival ferry operators on the route. Then, in 1986, after generations of aborted attempts to initiate a tunnel project, and in the teeth of bitter opposition from the ferry operators, the French and British governments finally gave the go-ahead to build the Channel Tunnel. In the face of these factors, the board rooms of ferry operators had to consider how best to conduct their prime function: to maximise a yield for their investors. It was against this background, that P & O Ferries had to make some bold business decisions. It was the management of the financial risks and opportunities which were uppermost among the concerns in the board room of P & O Ferries; safety management did not feature in their job descriptions. Out of that management of risk, they found a solution, when they resolved to buy out their very formidable rival, Townsend Thoresen, reducing the competition and spreading the overheads. In February of 1987, the deal was completed and the directors of P & O Ferries implemented their plans for maximising the financial yield.

P & O's own ferries on the Dover-Calais route were doing their job very satisfactorily; they did not need Townsend's tonnage there, so it was decided to switch the *Herald of Free Enterprise* to the Zeebrugge route. The only problem was that the linkspan at Zeebrugge had not been designed with the *Herald* in mind, so that, for the vessel's upper vehicle deck to be accessed by the ramp, it was necessary to trim the ship by the head and flood her ballast tanks, to lower the level of the vehicle deck to the linkspan.

When the *Herald* left Zeebrugge on 6 March 1987, not all the water had been pumped out of the bow ballast tanks, causing her to be some 3 feet down at the bow. Mr Stanley, the assistant bosun, was responsible for closing the bow doors but he had been released from duties by the bosun before the sailing time. He duly went to his cabin and fell asleep; tragically, he slept through the 'Harbour Stations' call which ordered the crew to their assigned sailing positions. It was not part of anybody else's duties to ensure that the bow doors were closed before sailing, save the duty of the Master to ensure that the vessel was in all respects safe to proceed to sea. Her design of clamshell bow doors made it impossible for Captain David Lewry to see from the bridge if the doors were opened or closed, though.

The *Herald* sailed at 19.05 local time, with a crew of 80 and some 459 passengers, 81 cars, three buses and 47 trucks. Passing the outer mole 19 minutes later, she increased speed, when a bow wave began to build up under her prow. At 15 knots, with the bow down 3 feet lower than normal, water began to break over the main car deck through the open doors at the rate of 200 tons per minute.

In common with other roll on – roll off vessels, the *Herald*'s main vehicle deck lacked transverse bulkheads and, so, the sudden flood of water through the bow doors quickly caused the vessel to become unstable. The *Herald* listed 30 degrees to port almost instantaneously, as water continued to pour in and fill the port wing of the vehicle deck, causing her to capsize 40 seconds later. The *Herald* settled on the sea bed at slightly more than 90 degrees with the starboard half of her hull above water. There had been no chance to launch any of the ship's lifeboats.

At least 150 passengers and 38 members of the crew lost their lives when the vessel capsized, the worst disaster for a British vessel in peacetime since the sinking of the *Titanic* in 1912.

The Inquiry

The prevailing legislation at the time for the conduct of an inquiry was the Merchant Shipping Act 1970, defining the statutory provision for a judicial investigation into the loss or presumed loss of a UK-registered ship anywhere in the world, as is the responsibility of the Flag State, or such loss of any ship that foundered in UK territorial waters, as is the responsibility of the Port State. This Formal Investigation was conducted by the Honourable Mr Justice Sheen,[73] whose findings sent the industry into nervous shock.

Precisely as we have seen in civil aviation cases, the purpose of the investigation was to establish findings on the loss of the vessel, with a view to preventing such casualties in the future. It was not the function of the process to establish criminal or civil liability; but the factual evidence heard in the Inquiry might have been used in litigation to prove – or discredit – such a claim. Mr Justice Sheen, the Admiralty Judge who was appointed Wreck Commissioner for the purpose of presiding over the judicial investigation, summarised his conclusion:

> The Court, having carefully inquired into the circumstances attending the above-mentioned shipping casualty, finds, for the reasons stated in the Report, that the capsizing of the Herald Of Free Enterprise was partly caused or contributed to by serious negligence in the discharge of their duties by Captain David Lewry (Master), Mr Leslie Sabel (Chief Officer) and Mr Mark Victor Stanley (Assistant bosun), and partly caused or contributed to by the fault of Townsend Car Ferries Limited (the Owners).

The Court had no powers of punishment and save the authority to suspend certificates of competence on findings of poor seamanship, and thus suspended *the certificate of the said Captain David Lewry for a period of one year from the 24th July 1987. The Court suspends the certificate of the said Mr Leslie Sabel for a period of two years from the 24th July 1987.*

The Inquiry heard the evidence of a director of the Company, Mr Develin, who, when asked who was responsible for considering matters relating to safety in the navigation of the Company's ships, answered, *Ashore, the system would be to take a consensus of the senior masters.* The Judge took the view that, without a properly qualified marine superintendent, it sounded sensible to rely on such a consensus; the problem revealed by the evidence, though, was that the shore management took very little notice of what they were told by their Masters.

The inquiry could not avoid drawing the conclusion, after analysing the exhaustive evidence, that the underlying fault lay within the management control structure, and those responsible for safety as a strategic function of the company. Some safety system had to be rolled out on a company-wide basis, which was down to the responsibility of the company.

In the Judge's opinion, the Company's Board of Directors' ultimate responsibility was to ensure that the safety of their passengers was maintained, which, to the Judge's mind, should have led them to ask themselves, what orders should be given for the safety of their ships? The Board, however, did not appear to have appreciated their responsibility for the safe management of their ships. As a result, not a single director had been charged with the duty of organising safety on the *Herald*. In terms of Board responsibility, this was understandable, if inexcusable – as with most companies, the Board's management function was designed primarily to ensure a profitable yield for the shareholders. Mr Justice Sheen's Report summarised his conclusions on the priority of commercial demands in the Pressure to leave the berth. The Court was left in no doubt that deck officers were under pressure that no time was to be wasted. The Company sought to say that this disaster could have been avoided if the Chief Officer had waited on G deck another three minutes. That is true. But the Company took no proper steps to ensure that the Chief Officer remained on G deck until the bow doors were closed. On 6 March, they were running late. The *Herald* sailed five minutes late. This may have contributed to the Chief Officer's decision to leave G deck before the arrival of Mr Stanley, which he anticipated.

This was one of the factors which led to the Judge's comments in the Report, which have subsequently rumbled thunderously through the boardrooms of every shipowner in the country:

> A full investigation into the circumstances of the disaster leads inexorably to the conclusion that the underlying or cardinal faults lay higher up in the Company. The Board of Directors did not appreciate their responsibility for the safe management of their ships. They did not apply their minds to the question: What orders should be given for the safety of our ships?
>
> The directors did not have any proper comprehension of what their duties were. There appears to have been a lack of thought about the way in which the Herald ought to have been organized for the Dover-Zeebrugge run. All concerned in management, from the members of the Board of Directors down to the junior superintendents, were guilty of fault in that all must be regarded as sharing responsibility for the failure of management. From top to bottom the body corporate was infected with the disease of sloppiness.[74]

The Consequence of the Case in Law

While Mr Justice Sheen had exercised his power to deal with the Master by way of reviewing his certificate of competency, the Judge had no power to order any redress against the Company; recourse would have to follow in a more appropriate Court. And while the Judge criticised the directors in the *Herald* Inquiry in the most scathing terms possible, the same words, paradoxically, became the directors' salvation. The fact that the evidence led to the conclusion that the directors had not appreciated their responsibility for the safe management of their ships, that they had not applied their minds to the question of safety and their lack of comprehension of what their duties were, showed that no director had assumed any personal responsibility which underpinned a duty of care and, in the absence of that, there could not be any realistic prospect of a conviction for manslaughter against an individual director, and thus, under the identification doctrine, no prospect of a successful Prosecution against the Company for corporate manslaughter.

Mr Justice Taylor had to wrestle with the concept of the controlling mind in the ensuing Prosecution of P & O Ferries in the *Herald* case,[75] when he re-stated the law that a corporation may be culpable of the crime of corporate manslaughter where an individual, who is part of the corporation's controlling mind, does an act which fulfils the pre-requisites for the crime of manslaughter. However, seriously the disease of sloppiness had taken hold, the mere failure of the management system to prevent a death, in the absence of an individual guilty of manslaughter, could not, itself, sustain a conviction. In the Trial against P & O, therefore, the Prosecution was doomed to fail. After a submission by the Defence at the end of the Prosecution case, the Judge ruled that the Prosecution was not in a position to satisfy the doctrine of identification.

Corporate accountability is seen by Board rooms as a threat in English law, with the development of the modern law on corporate manslaughter, inevitably precipitated by the failure of a successful Prosecution against P & O Ferries following the *Herald of Free Enterprise* casualty. In the light of the Law Commission's report in 2000,[76] the Select Committee on Home Affairs and Work and Pensions set out new proposals, intended to change the basis of liability from the requirement of identifying the causal link from an individual Guilty of manslaughter to the controlling mind of the company, to liability founded on accountability for the way in which an organisation's activities are managed or organised by its senior managers. In this way, it was the intention of Parliament to close down the possibility that a shipowner could avoid criminal liability for corporate manslaughter, in the way encountered in the case of the *Herald*.

THE 2007 ACT

The Corporate Manslaughter and Corporate Homicide Act 2007 was the result. The crime of corporate manslaughter is, broadly, committed when an organisation owes a duty to take reasonable care for a person's safety and the way in which activities of the organisation have been managed or organised amounts to a gross breach of that duty and causes the person's death. How the activities were managed or organised by senior management must be a substantial element of the gross breach, so the management function demands a relevant duty of care to the victim according to the common law of negligence.

As if to remove any lingering doubt or loophole in the application of the 2007 Act, Section 28 extends its territorial application to harm resulting in death in the UK or on a British-controlled aircraft as defined in Section 92 Civil Aviation Act 1982, even if the harm was sustained on an aircraft by a person who is then no longer on board the aircraft if it has been wrecked. Overall, the government servants made a thorough job of ensuring that the act would be applied vigorously to the aviation scenario, whether the incident takes place on an aerodrome or in the air; and there is no doubt that companies are understandably shy of exposing themselves any more than is absolutely essential to such a risk. The conduct of the Pilot in command or air traffic controller is not necessarily fatal to the company's position, but the potential for criminal litigation is obvious.

For the aircraft operator or air traffic control supplier, the serious issue is that there need no longer be a causal link between the fatality and the controlling mind of the company; now the professional's behaviour may incriminate them because the death must have been caused merely by a management failure; it need not even have been the sole cause of death, but if the professional, who, after all, plays significant strategic or regulatory compliance roles in the management of the whole or a substantial part of the organisation's activities, behaves in such a way that their conduct falls far below what could reasonably have been expected, then the corporate owner or operator will be guilty of the offence.

The offence is described in Section 1 of the Corporate Manslaughter and Corporate Homicide Act 2007, by which a company is guilty of an offence if the way in which its activities are managed or organised causes a person's death, and amounts to a gross breach of a relevant duty of care owed by the company. But the company is guilty of an offence only if the way in which its activities are managed or organised by its senior management is a substantial element of that breach. For the purposes of this act, a *relevant duty of care* means any of the duties owed by it *under the law of negligence* under the provisions on negligence in civil law. Given this, it is unclear why Parliament steered away from the director's duty of care, skill and diligence sensibly defined under Section 174 Companies Act, passed just one year before, which requires the evidential burden to be satisfied as to whether there was negligence in the understanding of civil law and its associated burden of proof; instead it applied a definition in terms of what falls far below the reasonable expectation of the company to meet Health and Safety statutory criteria for a duty – which, as strict liability offences, need no further evidence as to the standard reasonably to be expected in their discharge.

It is important to emphasise the point that a breach of a duty of care by a company is a 'gross' breach if the conduct which allegedly amounted to a breach of that duty falls far below what can reasonably be expected of the company in the circumstances; 'senior management' means the persons who play significant roles in the making of decisions about how the whole or a substantial part of its activities are managed or organised. In this respect, one might be forgiven for thinking that the identification doctrine has not really gone away at all; it will still be easier to prosecute a company with just one director, than with a whole host of 'senior managers' whose individual parts to play in the management activities must be analysed.

The civil aviation industry is extremely sensitive to security threats and acts of terrorism; examples of airport restrictions in security are ubiquitous, while the in-flight dangers illustrated by the attacks on Pan Am Flight 103 and Malaysian Airlines MH17 are readily apparent. Security Management Systems provide a formalised, risk-driven framework which is designed to embed security into the daily operations and culture of the airport or operator, enabling them to identify and address security threats, risks, gaps and weaknesses in a consistent and proactive way. If they have a system which contains all the elements that are identified in this framework it will certainly assist in complying with the UK quality assurance obligations under the CAA's basic level of regulatory requirements. The CAA's guidance is specifically aimed at the company's senior management personnel with an overview of the principles of a Security Management System, and recommends that each corporate operator should appoint an accountable manager who is a senior executive with access to the Board in order to influence decision making. The guidance offers some insight into the CAA's understanding of the 2007 Act insofar as it recommends that they hold the authority to allocate necessary resources for the protection of the public, staff, assets and the business itself.[77] The ultimate value of the system to the manager lies in its evidence trail, so that if there were an allegation of violating the 2007 Act, demonstration of compliance with the system would offer evidence that the company had complied with the requirements that met the minimum standard of responsibility; while the absence of such a system would amount to prima facie evidence of a management failure, that would have to be rebutted by evidence to the contrary.

Case Study: Case of Mark Forkner and the Boeing 737 Max

The Boeing 737 is a short- to medium-range twinjet narrow-body airliner developed and manufactured by Boeing Commercial Airplanes, part of The Boeing Company in the United States. The 737 was originally conceived in 1964 as a smaller, lower-cost twin-engine airliner derived from the 707 and 727, but its popularity was built on an ever-growing foundation of reliability and economic operation, making the 737 series the highest-selling commercial jetliner in history. Such a superlative must always be under review, of course, and the 737 was revised again in the twenty-first century for greater efficiency, with the 737 MAX series enhanced, particularly, with new engines, that had to be mounted further forwards and higher on the wing of the original airframe in order to maintain clearance from the ground; this resulted in the aircraft tending to pitch up more than previous models. As a result, Boeing developed the Maneuvering Characteristics Augmentation System (MCAS), which automatically pitched the nose of the plane down using the elevator trim to prevent the aircraft from entering a stall. Crews were not informed about MCAS in their training modules for the new jet, however and in an unprecedented cost-cutting move, the system relied on a single Angle-of-Attack indicator to operate, instead of having using to sensors to have one backup. The 737 MAX entered service in 2017 but, after a successful start, was grounded worldwide in March 2019 following two fatal casualties.

Lion Air Flight 610

On 29 October 2018, a 737 MAX 8, on an internal flight from Jakarta to Pangkal Pinang, Indonesia, crashed into the sea 13 minutes after take-off, with 189 people on board the aircraft, including the flight crew. There were no survivors.

NTSC Indonesia published a preliminary report on the Lion Air B737 MAX 8 accident,[78] On the day before, during a pre-flight check, the Pilot in command discussed with the company's maintenance engineer actions that had been performed including replacement of the Angle-of-Attack sensor, which had been tested accordingly. The flight then departed.

Subsequently, the Digital Flight Data Recorder showed the stick shaker activated during the rotation and remained active throughout the flight. At about 400 feet, the Pilot in command noticed on the primary flight display an Indicated Air Speed Disagree warning. The Pilot handed over control to the second in command and cross checked the primary flight displays with the standby instrument and determined that the left primary flight display had the problem, and he noticed the aircraft was trimming with the automatic nose down. He moved the Stab Trim switches to Cut Out and the second in command continued the flight with manual trim.

The Pilot declared *PAN PAN* to the Denpasar Approach controller due to instrument failure and requested to maintain runway heading. He performed three Non-Normal Checklists and none contained the instruction *Plan to land at the nearest suitable airport*. The flight eventually landed successfully at Jakarta. After parking, the Pilot informed the engineer about the problems encountered. The engineer duly investigated and the subsequent test on the ground found the problem to have been solved.

At 23.20, on 29 October, the aircraft departed from Jakarta with a flight plan for Pangkal Pinang. The flight recorder subsequently indicated a difference between left and right Angle-of-Attack readings of about 20 degrees and continued until the end of recording. During rotation, the left control column stick shaker activated and continued for most of the flight. The second in command sought independent confirmation of their position and reported to the air traffic controller a *flight control problem*, asking them to confirm the altitude of the aircraft and later also asked the speed as shown on the controller radar display. The flight recorder reported that, after the flaps retracted, the automatic nose down trim activated but stopped when the flaps extended. When the flaps were retracted to zero, the automatic trim began again and continued for the remainder of the flight. The last radio contact was made 11 minutes after take-off, when Pilot in command requested the arrival controller to block altitude 3,000 feet above and below for traffic avoidance. The controller asked what altitude the Pilot wanted, to which the captain answered, *Five thou*. Twenty seconds later, the flight recorder stopped recording.

Ethiopian Airlines Flight ET302

On 10 March 2019, a 737 MAX 8 on a flight from Addis Ababa, Ethiopia, to Nairobi, Kenya, crashed six minutes after take-off, killing all 157 people aboard including the flight crew. The aircraft was only four months old at the time of the accident. In rapid response, numerous aviation authorities around the world grounded the 737 MAX, and many airlines followed suit on a voluntary basis. On 13 March 2019, the FAA became the last authority to ground the aircraft, reversing its previous stance that the MAX was safe to fly.

The take-off roll had appeared normal, including normal values of left and right Angle-of-Attack, but shortly after rotation, the left and right recorded Angle-of-Attack values deviated and then responded erratically. At the same time, the left stick shaker activated and remained active, while the airspeed, altitude and flight director pitch bar values from the left side noted deviating from the corresponding right side values. The Pilot in command attempted to engage the autopilot twice, but this resulted in two autopilot warnings. He instructed his second in command to contact the radar controller to warn them, when they reported another flight departure climbing to a flight level of 32,000 feet. The avionic systems continued to behave erratically, and the aircraft responded with roll oscillations accompanied by lateral acceleration, rudder oscillations and slight heading changes.

Within a minute, the air traffic controller instructed them to climb to 3,400 feet and, when able, to make a right turn direct to an established navigation waypoint near Lake Turkana, on the Ethiopia-Kenya border. The final events were summarised in the accident report,[79] from which the following has been epitomised.

After the flaps were fully retracted, the first automatic nose-down trim activated and engaged for nine seconds positioning the stabiliser trim to 2.1 units. The Pilot pulled to pitch up the aircraft with a force more than 90 lbs. He then applied electric trim-up inputs. Five seconds after the end of these inputs, a second automatic nose-down trim activated. Following nose-down trim activation, a ground proximity warning system sounded for three seconds and *Pull Up* displayed on the primary flight display for three seconds. Approximately five seconds after the end of the crew manual electrical trim up inputs, a third automatic trim nose-down was recorded but with no associated movement of the stabiliser.

The Pilot in command told his co-pilot, *Advise we would like to maintain one four thousand. We have a flight control problem.* Air Traffic Control duly approved. The new target altitude of 14,000 feet was duly set, but the Pilot was unable to maintain the flight path and requested to return back to the departure airport, but the avionics continued to respond erratically and in less than four seconds, the pitch angle dropped from 0.5 degree nose up to 7.8 degrees nose down, as the descent rate increased from 100 feet per minute to more than 5,000 feet per minute. At the end of the flight, computed airspeed values reached 500 knots, pitch values were greater than 40 degrees nose down and descent rate values were greater than 33,000 feet per minute. Both flight recorders stopped recording three minutes later, at 05.43.

THE TRIAL OF MARK FORKNER

Investigations in both crashes determined that Boeing had made cost-saving solutions to defects which had been reported, but the solution of the MCAS was fatally flawed. Additionally, the investigations revealed that Boeing had concealed highly probative evidence in order to obtain due certification in relation to the MCAS.

In the United States, there is no current law of corporate manslaughter such as that in the UK and European States. However, if an employee commits a crime within the sphere of their employment and with the intention of benefitting the corporation, their guilt can be imputed to the company under the doctrine of vicarious liability. In this respect there was a very realistic possibility that Boeing might face corporate accountability arising out of the fatal accidents. The issue arose out of an allegation that a Boeing employee, Mark Forkner, had dishonestly deceived the Federal Aviation Administration (FAA) during its evaluation of Boeing's 737 MAX.

Forkner was Boeing's 737 MAX chief technical pilot, with responsibility for leading the process to provide the FAA with true, accurate and complete information for its safety evaluation. It was alleged that, in or about November 2016, Forkner had discovered information about an important change to MCAS. Rather than sharing information about this change with the FAA, he allegedly intentionally withheld this information and deceived them about the safety reliability of MCAS, with the intention of obtaining a positive safety evaluation on an unsafe aircraft. As a result, Pilots flying the 737 MAX were not provided with any information about MCAS in their manuals and training materials. After the FAA were appraised of the Lion Air crash, they discovered the information about the important change to MCAS that Forkner had withheld. Having discovered this information, the FAA AEG began reviewing and evaluating MCAS but, in the meantime, they learned of the Ethiopian Airlines crash.

The US Department of Justice charged Forkner with two counts of fraud involving aircraft parts in inter-State commerce and four counts of wire fraud, dishonestly making use of communication technology. He was duly arraigned on a long indictment[80] and, if found Guilty, might have been sentenced to a maximum of 20 years in prison on each of four counts. He pleaded Not Guilty. However, the Prosecution of Mark Forkner ended in acquittal by a very rapid Jury verdict, raising questions about the fundamental failure of the Department's pursuit of a relatively middle-ranking employee while allowing senior officers and board members to escape accountability. Inevitably, unconfirmed suspicions were aroused that the Department of Justice had trod softly with the Boeing corporation when it became apparent that Forkner had been the only Boeing employee charged with a crime; indeed, in 2021, the Justice Department said the misconduct at Boeing was not facilitated by senior management.

But worse was in store for Boeing.

UNITED STATES V THE BOEING COMPANY[81]

In 2021, Boeing was charged with fraud conspiracy, on the basis of vicarious liability that Forkner had provided the FAA Aircraft Evaluation Group with materially false, inaccurate and incomplete information about the Maneuvering Characteristics Augmentation System. Boeing entered into a Deferred Prosecution Agreement (DPA) with the Fraud Section of the Department of Justice and the US Attorney's Office for the Northern District of Texas to resolve the charge of conspiracy to defraud the Federal Aviation Administration's Aircraft Evaluation Group in connection with the safety evaluation. Boeing admitted in court documents that two of its 737 MAX Flight Technical Pilots deceived the FAA AEG (although only Forkner was arraigned), but a long and expensive Court battle would have ensued to establish its vicarious liability. In due settlement, under the DPA agreed to pay a total of over $2.5 billion, consisting of a criminal penalty of $243.6 million, compensation payments to Boeing's 737 MAX airline customers of $1.77 billion, and a $500 million fund (Fund) to compensate the heirs, relatives, and legal beneficiaries of 346 individuals who died in the crashes. The DPA required Boeing to set up a claims-handling system to distribute the fund and to bear all the costs associated with that process.

The DPA was scheduled to last for three years, during which Boeing was banned from making any public statement, in litigation or otherwise, contradicting its acceptance of responsibility and admission of the facts set forth in the DPA's Statement of Facts. Boeing also agreed not to use the fact of any of the payments required by the DPA, including the Fund, to prevent the Fund's beneficiaries from pursuing any other lawful claim that they might have against Boeing. Boeing also agreed that it would be held in breach of the DPA should it commit any US federal felony during the term of the DPA, provide any deliberately false, incomplete or misleading information in connection with the DPA, fail to fully cooperate as required by the DPA, fail to implement a compliance programme as required by the DPA or fail to completely perform or fulfil each of their obligations under the DPA. Should Boeing breach any of the conditions of the DPA, the Department of Justice would have the power to prosecute Boeing for the crime charged in the information.

Still that was not the end. Families of the victims commenced proceedings for compensation under the Crime Victims' Rights Act,[82] in which they argued that they were excluded from the negotiations for the DPA, compromising their rights under State law. The matter went to the US Court of Appeals for the Fifth Circuit. On 15 December 2023, the Court issued its decision, affirming the decision of the lower Court that it could not substantively revise the DPA between the government and Boeing. But

> if the Government concludes, independent of court supervision, that Boeing has not complied with the DPA, the case will instead proceed to Trial or to Rule 11 guilty plea resolution, both assuring CVRA victim protection. On the other hand, if the Government assesses that Boeing has complied sufficiently with the DPA's terms and asks the district court to dismiss criminal proceedings, then, 'in passing on any government motion under Rule 48(a) … the court will expect to see the prosecutor recount that the victim has been consulted on the dismissal and what the victim's views were on the matter.

In January 2024, Boeing's risk management faced the crisis of a lucky escape from disaster when a defective door plug on an Alaska Airlines 737 MAX forced an emergency landing as the escape hatch was ripped off in mid-flight. The company's share value suffered dramatically, in what, ultimately, could precipitate a problem worse than a criminal conviction: shareholder flight.

SENTENCING

While the strict liability offences provided in civil aviation legislation clearly are intended to deliver redress in a cost-effective process, the way in which that redress is carried out is a matter for the judicial processes of the individual States which have ratified the Chicago Convention – and they may go far beyond the Convention limitations of a fine. In the UK, the Sentencing Guidelines Council published a helpful paper in 2004, addressing overarching and general principles relating to the sentencing of offenders. In 2018 new, additional guidelines were issued to Judges which increase prison sentences for Defendants convicted of gross negligence manslaughter in a workplace setting – which includes the civil aviation setting.[83]

Criminal procedure broadly compels the Court to take into account five purposes of sentencing when dealing with convicted offenders: the punishment of offenders, the reduction of crime (including its reduction by deterrence), the reform and rehabilitation of offenders, the protection of the public and the making of reparation by offenders to persons affected by their offence. The application of those purposes in the context of the aviation professional in fact underpins the relevance of all save the reform of offenders, which is not a matter with which the prosecuting authority is likely to concern itself.

The overarching obligation on the Court is to pass a sentence that is fair and proportionate when considered with the seriousness of the offence. The seriousness of an offence is determined by two main parameters: the culpability of the offender and the harm caused or risked being caused by the offence.

CULPABILITY

Culpability, or blameworthiness on the part of the offender, will be determined by such factors as motivation, whether the offence was planned or spontaneous or whether the offender was in a position of trust – or responsibility. Such features manifestly lend themselves to considering the position of a seafarer in a casualty event and, thus, the relevance to the task in context is compelling. Sentencing in cases of gross negligence manslaughter follows Section 125(1) of the Coroners and Justice Act 2009, which provides that every Court must consider the appropriate sentencing guidelines, even though the factors indicate a lower culpability, where there was no intention by the Defendant to cause any harm and no obvious risk of anything more than minor harm. This would most likely address the category into which the aviation professional would fall.

In cases of non-fatal accidents, such as those encountered in runway incursions or air traffic failures, culpability has been analysed by the Council to have four levels, which determine the seriousness attached to the *mens rea* of the offender and can be paraphrased to apply in seafarers as follows:

- The Defendant had the intention to cause harm, with the highest culpability when the offence itself was planned. The worse the harm intended, the greater the seriousness.
- The Defendant was reckless as to whether harm would be caused, that is, where they appreciated that at least some harm would be caused but proceeded any way, giving no thought to the consequences even though the extent of the risk would be obvious to most seafarers in that position.
- The Defendant had knowledge of the specific risks entailed by their actions even though they did not intend to cause the harm that resulted.
- The Defendant was guilty of negligence. (Sadly, the word 'negligence' is not defined by the Council.)

HARM

Harm must always be judged in light of, and, therefore, must follow the determination of, culpability. The Council has analysed the parameter of harm to cover three types:

- Harm to the individual, who may suffer death, personal injury, psychological stress or financial loss; in other words, issues which are personal to the victim in question, and cause a particular impact which the Court should reasonably take into account when determining the seriousness of the offence.

- Harm to the community, which widens the impact from the individual considered above to the well-being or otherwise of the local Society, or Society at large.
- Other types of harm, which the Council had difficulty in characterising but, in the context of an aviation casualty, may describe downstream consequences arising out of the event, such as fleet groundings or other disruption as a result of findings of investigation reports.

The Council acknowledged that some conduct is criminalised purely by the normative ethics of Society, in particular when public feeling is inflamed by the consequences of the behaviour, which can influence public perception of the harm caused.

AVIATION SECURITY

Part I of the Aviation Security Act 1982 contains penalties for offences endangering the safety of aircraft, and provides for maximum penalties. The Air Navigation Order 2016, made under Section 61 Civil Aviation Act 1982, creates a number of offences designed to secure the safety of civil aircraft, some relate to the conduct of passengers and air traffic controllers as well as to aircrew. The decision to prosecute will almost certainly be in the public interest, given the following:

- public safety has been placed at risk;
- passengers or staff have sustained loss, damage or personal injury;
- serious or widespread disruption and inconvenience has been caused to persons using the transport system.

Not so clearly will it be in the public interest in cases in which there is a purely technical breach of the law provided that:

- there has been no risk to public safety;
- the offence resulted from a genuine oversight or misunderstanding; and
- no Injury or loss has been sustained by either passengers or staff.

Case Study: Roderick Brown

This recent case illustrates the wide-ranging application of aviation security.

All drivers transporting secure air cargo must be trained by an Aviation Security Certificated Instructor. The Defendant was an aviation security training instructor who was alleged to have falsified training certificates and training records for drivers working in the air cargo security sector, following an investigation led by the UK CAA. In December 2023, at the Crown Court in Lewes, he pleaded Guilty to four counts of forgery and was sentenced to 30 weeks imprisonment, suspended for 18 months, with a condition to perform 100 hours unpaid work. He was also ordered to pay costs of £750 to the CAA who brought the Prosecution. Peter Drissell, director of Aviation Security for the CAA, commented:

> The effectiveness of aviation security relies on ensuring that appropriate training is delivered to staff across the industry by qualified instructors via registered training providers. The UK Civil Aviation Authority auditors uncovered a clear abuse of trust and responsibility on the part of an individual in this case, and the decision to prosecute was taken to protect the public and professional confidence in the high standards we set for aviation security in the UK.[84]

NOTES

1 Ormerod, D, 2008, *Smith & Hogan Criminal Law*, Oxford University Press, Oxford, p. 7.
2 Ormerod, D, 2008, *Smith & Hogan Criminal Law*, Oxford University Press, Oxford, p. 7.
3 Baker, J, 2007, The Moral Limits of Criminalizing Remote Harms. *New Criminal Law Review* 11(3) (2007) 371.
4 Ericson, R and K Carriere, 1994, The Fragmentation of Criminology, in D Nelken (ed), *The Futures of Criminology*, Sage, London, p. 89.
5 See *R v Moloney* [1985] 1 All ER 1025; *R v Nedrick* [1986] 3 All ER 1.
6 Mukherjee, P, 2006, Criminalisation and Unfair Treatment: The Seafarer's Perspective. *Journal of International Maritime Law*, 12 (2006) 325–336.
7 *R v G* [2004] 1 AC 1034. See Richardson, P (ed), 2009, *Archbold Criminal Pleading, Evidence and Practice*, Thomson Reuters (Legal) Limited, London, p. 1855.
8 Regulation (EU) No. 996/2010. Updated 15 January 2021, and adopted by the UK.
9 Annex 13, para 2.2.
10 Annex 13, para 2.3.
11 Art 12.
12 *R v Deller* [1952] 36 Cr App Rep 184.
13 It did not do Mr Deller a lot of good in the end, for the Court held that he could now be convicted of an *attempt* to obtain property by deception, which did not require the final element of the *actus reus*. See Ormerod, D, 2008, *Smith & Hogan Criminal Law*, Oxford University Press, Oxford.
14 Actus non facit reum nisi mens sit rea.
15 *Sweet v Parsley* [1970] AC 132, at p. 148.
16 *R v Prince* [1875] LR 2 CCR 154.
17 See *R v Caldwell* [1981] 1 All ER 961; *R v Lawrence* [1981] 1 All ER 974.
18 *R v Henderson*, 2021, Cardiff Crown Court (unreported).
19 See the Judge's sentencing remarks: www.judiciary.uk/wp-content/uploads/2022/07/R-v-Henderson-Sentencing-Remarks-12.11.21.pdf.
20 See mv *Herald of Free Enterprise*. Report of Court No. 8074. Formal Investigation (July 1987).
21 *R v P & O European Ferries (Dover) Ltd* [1991] 93 Cr App R 72.
22 *Donoghue (or McAlister) v Stevenson* [1932] All ER Rep 1; [1932] AC 562.
23 *Donoghue (or McAlister) v Stevenson*, at p. 580.
24 *The Argonaftis* [1989] 2 Ll R 487.
25 *Caparo Industries plc v Dickman* [1990] 2 AC 605; [1990] [1990] UKHL 2; subsequently applied in *Marc Rich & Co AG and Others v Bishop Rock Marine Co Ltd and Others* [1995] 3 All ER 307; [1995] UKHL 4; [1996] 1 AC 211; [1995] CLC 934; [1995]

2 LLR 299; [1996] ECC 120; [1995] 3 WLR 227; [1995] 2 Lloyd's Rep 299, in which Lord Steyn said that the common law of negligence develops incrementally on the basis of a consideration of analogous cases where a duty has been recognised or desired.
26 *R v Evans* [2009] EWCA Crim 650.
27 A statement 'obiter' is one made by a Judge which is not essential to the case in hand but goes further in explaining the general principle of law involved.
28 Michaelides-Mateou, S and A Mateou, *Flying in the Face of Criminalization: The Safety Implications of Prosecuting Aviation Professionals for Accidents*, Routledge, Abingdon.
29 See below.
30 *Lister and Others (AP) v Hesley Hall Limited* [2001] UKHL 22.
31 Tadros, V, 2006, 69(4) MLR 601-618, at p. 601.
32 *Andrews v DPP* [1937] 2 All ER 552, at pp. 554–555.
33 *R v Adomako* [1995] 1 AC 171.
34 *R v Bateman* [1925] 19 Cr App R 8, at pp. 12–13.
35 *R v Adomako* [1995] 1 AC 171.
36 In addition, the breach of duty was 'a substantial' cause of death; as refined by *R v O'Connor* [1997] Crim LR 16 (CA).
37 In this case, meaning the Jury's conclusion.
38 *R v Adomako* [1995] 1 AC 171.
39 *R v HM Coroner for Inner London, ex parte Douglas-Williams* [1998] 1 All ER 344.
40 *R v Cheshire* [1991] 1 WLR 844.
41 *R v Joanne Rudling* [2016] EWCA Crim 741 at para 18.
42 *Honey Maria Rose v R* [2017] EWCA Crim 1168.
43 See www.raf.mod.uk/bbmf/theaircraft/dakotahistory.
44 NTSB Aviation Investigation Final Report accident number: CHI97FA084.
45 *Camarata Property Inc v Credit Suisse Securities (Europe) Ltd* [2011] EWHC 479 (Comm).
46 *R v Misra and Srivastava* [2004] EWCA Crim 2375.
47 *Haughton v Smith* [1975] AC 467.
48 *Caparo v Dickman*, supra.
49 *R v Adomako* [1995] 1 AC 171.
50 *Williams and Another v Natural Life Health Foods Ltd* [1998] 1 WLR 830.
51 *R v K* [2001] UKHL 41.
52 See *R v Williams (Gladstone)* [1987] 3 All ER 411, 415.
53 *R v G and another* [2003] UKHL 50.
54 *Nam si causae ipsae pro se loqui possent, nemo adhiberet oratorem* – MT Cicero, Pro A Cluentio. 139.
55 *R v G* [2004] 1 AC 1034.
56 Such a person may fairly be accused of stupidity or lack of imagination, but neither of those failings should expose him to conviction of serious crime or the risk of punishment.
57 *R v Sellu* [2016] EWCA Crim 1716.
58 *Honey Maria Rose v R* [2017] EWCA Crim 1168.
59 *Andrews v DPP* [1937] 2 All ER 552, at pp. 554–555.
60 Convention for the Unification of Certain Rules for International Carriage by Air 1999.
61 See www.casa.gov.au/rules/compliance-and-enforcement/strict-liability.
62 The Law Commission, 2010, Consultation Paper No. 195 (Overview), *Criminal Liability in Regulatory Contexts.*
63 Kun, L, May 2008, Professional Responsibility: Politics, Culture, and Religion versus Science and Technology. *IEEE Open Journal of Engineering in Medicine and Biology* 27 (3) 95–99. doi:10.1109/MEMB.2008.918695. ISSN 0739-5175. PMID 18519190.
64 Solbrekke TD and T Englund, November 2011, Bringing professional responsibility back in. *Studies in Higher Education* 36 (7) 847–861. doi:10.1080/03075079.2010.482205. ISSN 0307-5079. S2CID 144219656.

65 Lindgren, J, S Carlbaum, A Hult and C Segerholm, 10 December 2018, Professional Responsibility and Accountability?: Balancing Institutional Logics in the Enactment of New Regulations and Practices against Bullying and Degrading Treatment in Swedish Schools. *Nordic Studies in Education* 38 (4) 368–385. doi:10.18261/issn.1891-2018-04-06. ISSN 1891-5949. S2CID 158295321.
66 Demuijnck, G, P Murphy and Philosophy Documentation Center, 2022, Retail Practitioners' Views vs. Ideal Theoretical Positions Concerning Ethical Business Practices with Garment Suppliers. *Business and Professional Ethics Journal* 41 (1) 59–92. doi:10.5840/bpej20211230114. ISSN 0277-2027. S2CID 245927006.
67 The International Civil Aviation Organization website elaborates. See www.icao.int.
68 Eurocontrol, 2018, Model for a Policy Regarding Criminal Investigation and Prosecution of Aviation and Railway Incidents and Accidents.
69 *Donoghue (or McAlister) v Stevenson* [1932] All ER Rep 1; [1932] AC 562.
70 *HL Bolton (Engineering) Ltd v TJ Graham and Sons* [1957] 1 QB 15.
71 *Tesco Supermarkets Ltd v Nattrass* [1972] AC 153 [HL].
72 *R v P & O European ferries (Dover) Ltd* [1991] 9 Cr App R 72.
73 mv Herald of Free Enterprise. Report of Court No. 8074. Formal Investigation (July 1987).
74 mv Herald of Free Enterprise. Report of Court No. 8074. Formal Investigation (July 1987), para 14.1.
75 *R v P & O European Ferries (Dover) Ltd* [1991] 93 Cr App R 72.
76 Reforming the Law on Involuntary Manslaughter: The Government's Proposals (Home Office, May, 2000, CC NO77828).
77 CAA, 2021, Security Management Systems (SeMS), A guidance note for Accountable Mangers.
78 Komite Nasional Keselamatan Transportasi Republic of Indonesia, Aircraft Accident Investigation Report, Jakarta, November 2018.
79 The Federal Democratic Republic of Ethiopia Ministry of Transport and Logistics, Aircraft Accident Investigation Bureau Investigation Report number AI-01/19, 23 December 2022.
80 US Department of Justice. See www.44778/Downloads/mark_forkner_indictment_redacted.
81 The Boeing Company Deferred Prosecution Agreement (DPA) Court Docket No.: 4:21-CR-005-O (N.D. Texas), circulated by the US Department of Justice Criminal Division.
82 18 U.S.C. § 3771.
83 See in particular www.sentencingcouncil.org.uk/offences/crown-court.
84 See www.caa.co.uk/newsroom/news/aviation-security-training-instructor-prosecuted/.

2 What Is Criminalisation in Civil Aviation?

OVERVIEW

Criminalisation is defined as the process by which crimes are created which lead to the prosecution of offenders that previously were not considered crimes by act or omission, but which have been made crimes by Society notwithstanding the human rights of the offender.

This chapter explains the evolution of criminalisation, so that we can place it in the context of the risk management of the professional in civil aviation; in other words, who will take the blame?

In aviation, the most crucial issue is safety. Article 44 of the Chicago Convention of 1944 highlights the overall objective of International Civil Aviation Organisation to ensure the safe and orderly growth of international civil aviation throughout the world. By their definition, accidents are the result of an undesirable chain of events, so we are not dealing with accidents caused intentionally or recklessly, which are subject to well-established tenets of criminal law. To prevent the repetition of accidents, investigations demand an effective reporting system, involving the willing effort of participating aviation professionals. If they fear that they will be prosecuted for conduct that previously had not been considered crimes, naturally, the voluntary process of participation will be compromised.

As the process of criminalisation has been embraced globally, it has taken on a much wider application, meeting the demands of the Risk Society, in which Society has become more concerned with public safety than with issues of fairness under the normal process of the law. This concept of risk management by domestic societies in sovereign states characterises the approach of investigations and prosecutions in civil aviation cases, which has redefined the concept of accountability.

It is this accountability which puts them in the dock; in the Trial that follows, they can be held to have met their responsibilities and discharged them within the law of the State in which they are prosecuted, in which case they will be discharged; if they fail to discredit the evidence against them, however, they will be held liable, and sentenced according to the powers of the Court.

In Chapter 1, we discussed the defining points in criminal accountability according to the sources of law to be found in particular States. The basis of criminalisation goes a step further, defining violations of the law that previously were not considered

DOI: 10.1201/9781003489696-2

crimes, but which have been made crimes by Society notwithstanding the human rights of the offender. In a paper delivered in 2004, Professor Edgar Gold introduced the process of criminalisation[1] as a problem which had been emerging *for some time*. This reveals two assumptions, which must be tested in the context of the aviation professional:

- Have they undergone a process of criminalisation?
- Has this been an emerging problem for some time?

The professional conduct of Pilots has shared a great deal of history with ships' Masters, and we can learn much from both; the cases against Masters, however, have the advantage that there are more of them, because, of course, they survived the casualty to face criminalisation. The Master has always been subject to the Law of the Flag State; theirs is the ultimate authority for that vessel and they are responsible to the Flag State for compliance with its maritime regulations. While the Master retains absolute discretion, they remain accountable for their command decisions. But the Master had better have good evidence that they managed the risk to the expected standard. The Human Element has emerged by analysis of shipping disasters in recent years, producing an increasing awareness of the central importance of the human element, everyone involved in the shipping industry needs to understand that they, themselves, are the human element. The same applies in civil aviation.

In exactly the same way as the pilot in command, the Master has a duty to take the standard of care, reasonably expected of such a professional in their position, to guard against acts or omissions which might cause loss and give rise to a claim in negligence. But in what circumstances will their default amount to a crime? Thus, they are exposed to the phenomenon of accountability that takes us beyond civil liability to criminal negligence. In this respect, the Master's Absolute Discretion in the International Convention for the Safety of Life at Sea[2] comes very close in meaning and in form to the Standardised European Rules of the Air,[3] abbreviated to SERA.

SERA.2010 Responsibilities

(a) Responsibility of the pilot-in-command

> The pilot-in-command of an aircraft shall, whether manipulating the controls or not, be responsible for the operation of the aircraft in accordance with this Regulation, except that the pilot-in-command may depart from these rules in circumstances that render such departure absolutely necessary in the interests of safety.

Perhaps most critical to the authority of the Pilot in Command is the provision contained in Regulation (EU) No. 923/2012:

> SERA.2015 Authority of pilot-in-command of an aircraft
> The pilot-in-command of an aircraft shall have final authority as to the disposition of the aircraft while in command.

Against this principle of absolute discretion, must be set the consequences. The evolution of criminal accountability has emerged in recent years as Society had demanded a response from its lawmakers that met its normative ethics. It is the normative ethics of the Society of the enforcing state which determines how (to paraphrase WS Gilbert) the punishment should fit the crime. In understanding the mischief of criminalisation, therefore, it is essential to analyse how the normative ethics of Society have influenced the current law and influences its future trends. So where did it start?

Many aviation professionals with long memories will smile in associating the words of Basil Lubbock with the Pilot in Command, although he was speaking of the Masters who commanded in that sisterhood of the China tea clippers:

> There were a few men, who held the necessary qualities of a tea-ship commander, whose endurance equalled their energy, whose daring was tempered by a good judgment, whose business capabilities were on a par with their seamanship, and whose nerves were of cast iron.[4]

Well, maybe those words provoke memories that are not quite so distant in civil aviation, but the normative ethics of Society, naturally, evolve in response to the concept of a risk which might threaten the security of that Society. Ericson and Carriere have defined this in terms of what they label a 'risk Society,' in which Society has become more concerned with public safety than with the issues of economic inequality which traditionally dominate its ethics.[5]

It is all about risk management, the concept which the high contracting parties adopt in formulating transnational law, which is the source of civil aviation law that bonds individual sovereign states to each other. How they respond to evolving concepts that shape the criminalisation of the aviation professional plays a crucial part in the foundation of this work.

The approach taken to legislation by a risk Society presents a clear tension which has a dynamic on the definition of a moral wrong in the context of normative ethics. The underlying causes of this tension may be identified in terms of a concern, inter alia, for the safety of the individual, for Society and for the State.[6] Such conflicting issues in this tension compel the risk Society to practise the Judgment of Solomon, and trade off the merits of ethical arguments which may place the individual and the State on a collision course. For the judgment to demonstrate that it has met the fundamental philosophical demands of Justice, the risk Society has to make decisions which must strike a balance between the regulation of criminal activity by the restraint of punishment, and the respect and maintenance of limits on such punishment. Inevitably the clash comes when the judgment must confront the seemingly intractable problem of rationalising and justifying the argument for deciding what human rights should be suspended in the interests of the sovereign State in question. Put in context, this challenges Society to convey a compelling argument to make the Pilot in Command accountable for wrongs which the risk Society perceives it to have suffered – simply because of the fact of the Pilot's responsibility, rather than the presence of any characteristics of a crime.

In 1995, Bennun commented that the prosecution of a pilot was a rare event.[7] It still is, but, maybe, that is because pilots historically have not survived air crashes. The case of Captain Stewart provided an early example of a prosecution.

Case Study: Captain Stewart and November Oscar

In this case, the CAA brought a prosecution against Glen Stewart, pilot in command of a Boeing 747 who made a defective bad-weather approach while attempting to land at London's Heathrow Airport on a flight from Mauritius.

On 21 November 1989, the aircraft came extremely close to the ground at London Heathrow Airport in thick fog before reversing its descent on an unsuccessful instrument approach. While the flight was over Frankfurt, the crew had received word that the weather at Heathrow had deteriorated and they would probably receive a category III landing, meaning that the fog was so low that an exceptionally precise autopilot-controlled approach and landing would have to be performed in near zero-zero conditions. The co-pilot was not certified to participate in such an instrument approach, and the Sperry SPZ-1 dual autopilot system could not be trusted. Taking everything into account, the pilot in command preferred to undertake the landing manoeuvre using his own discretion and judgment, which nobody could override, in accordance with his rights and obligations in law. If he made the wrong decision, he would be criminally accountable for the consequences. The aircraft therefore attempted a Category II landing in thick fog at London's Heathrow Airport and deviated from the centre-line of the runway, forcing a go around after descending to 75 feet outside the airport fence, which was parallel to a busy road, and was flying just 5 feet higher than a nearby hotel. In the end, he landed safely with 255 passengers, but the recklessness of his decision in which he exercised his discretion was taken to the point of criminal accountability. Charges were brought against Captain Stewart under the equivalent provisions today of Articles 240 and 241 Air Navigation Order 2016:

> Endangering safety of an aircraft
>
> Article 240. A person must not recklessly or negligently act in a manner likely to endanger an aircraft, or any person in an aircraft.
>
> Endangering safety of any person or property
>
> Article 241. A person must not recklessly or negligently cause or permit an aircraft to endanger any person or property.

On 8 May 1991, a jury at the Crown Court at Isleworth reached a majority verdict of 10 to two, finding Stewart Guilty of recklessly endangering the aircraft and passengers[8] (today Article 240). He was found Not Guilty on the other charge, which alleged recklessly endangering people or property on the ground (241). There had been no precedent under English law for such criminal accountability, and he was fined £2,000 or 45 days imprisonment.

The case proved to be highly controversial and divided opinion among legal and aviation professionals on the merit of a prosecution in the first place. The Prosecution alleged that the pilot in command failed to initiate the go-around expeditiously, without compelling expert evidence in support, while the Defence relied on evidence of defective instrument landing systems equipment and the incapacity of the co-pilot due to sickness, not to mention the dwindling state of fuel, due to a holding instruction and the conduct of the air traffic controller, who withheld clearance for November Oscar to land until the very last moment, certainly later than regulations permitted, to accommodate an Air France flight still rolling down the runway. Moreover, while juries are notoriously capricious in their decisions, there is reason to believe that they simply did not understand the technical evidence upon which much of the Defence case relied. The Trial took 11 days of extremely complex, technical evidence, before a Judge who admitted that he was woefully ignorant of even basic technical facts upon which he could help them in his summing up. Therefore it was perhaps unsurprising that, eventually, they reached a majority verdict on one charge and an acquittal on the other, without demonstrating clear understanding, but it is not open to question Juries on their decisions.

It would be the first case of criminalisation in English aviation law. It is noteworthy, however, that the law is limited not only to the pilot in command but also to any person who endangers the safety of an aircraft, as demonstrated six years later by a prosecution in the United States, against an aircraft's maintenance contractor. It is this latter scenario that we must address with a separate study.

Case Study: the Valujuet DC-9 crash[9]

Perhaps the first corporate criminal prosecution, on 11 May 1996, ValuJet Airlines McDonnell Douglas DC-9 was on Flight 592, a regularly scheduled flight from Miami to Atlanta. Shortly after 14.00, the aircraft departed and began a normal climb; but six minutes later, the pilots heard a loud bang in their headsets and noticed that the plane was losing electrical power. Seconds later, a flight attendant entered the cockpit and informed the flight crew of a fire in the passenger cabin, and passengers' shouts of "fire, fire, fire" were recorded on the cockpit voice recorder when the cockpit door was opened. The pilot in command immediately asked air traffic control to return to Miami and were given instructions for a return to the airport, but the aircraft was out of control. Eyewitnesses reported that the plane banked sharply, rolled onto its side and nosedived into the ground, when the aircraft was destroyed on impact. All on board were killed. The National Transportation Safety Board (NTSB) report[10] concluded that the fire had developed in a cargo compartment below the passenger cabin. The cargo compartment was designed to accomplish fire suppression by sealing the hold from outside air, so exhausting the fire of oxygen. The NTSB determined that just before departure, 144 expired chemical oxygen generators had been placed in the cargo compartment in five boxes marked COMAT (company material) by ValuJet's maintenance contractor SabreTech. This, however, violated Federal Aviation Administration regulations forbidding the transport of hazardous materials in passenger-aircraft cargo holds. In any event, SabreTech employees had recorded on the cargo manifest that the generators were empty. ValuJet workers then loaded the boxes in the cargo hold in the mistaken belief that the devices were simply empty canisters that would be safe and legal to transport on a passenger aircraft.

They were not empty, however. Investigators found that one of the oxygen generators was likely triggered when the plane experienced a slight jolt while taxiing. As the aircraft taxied and took off, the generator began releasing heat that caused other canisters to activate. Each activation created more heat, which rapidly caused all of the generators to activate. The intense heat ignited a fire in the other materials in the cargo hold. The fire was worsened by the presence in the cargo hold of two main aircraft tyres, which were also included in the list of materials shipped as COMAT. The tyres acted as an accelerant for the fire.

The NTSB concluded with responsibility on three parties:

- SabreTech, for improperly packaging and storing hazardous materials
- ValuJet, for not supervising SabreTech
- The FAA, for not mandating smoke-detection and fire-suppression systems in cargo holds as recommended in 1988 after a similar incident.

In 1997, a federal grand jury indicted SabreTech on 110 counts of manslaughter and third degree murder, as well as the violation of hazardous materials regulations, and two of SabreTech's mechanics were charged with conspiracy and making false statements. ValueJet was not prosecuted for the lack of supervision over their maintenance contractor.

Two engineers were acquitted but SabreTech were found guilty. In May 2001, SabreTech was fined $2 million. The fine was later reduced to $1.75 million, but, in November 2001, the United States 11th Circuit Court of Appeals overturned most of the verdict against SabreTech, on the basis that federal law did not support criminal penalties for reckless violations of regulations under the Hazardous Materials Transportation Act, which did not provide criminal penalties. The remaining items on the indictment were upheld, and SabreTech was fined $500,000 and three years' probation.

THE MISCHIEF OF CRIMINALISATION

In his paper on criminalisation, Professor Proshanto Mukherjee[11] succinctly summarised the characteristics of a crime in the context of the professional person with sharp focus on the *mens rea*, required in most serious crimes, which Society views as being moral wrongs in which the key element of intention or recklessness must be proved beyond reasonable doubt, according to the criminal burden. Failure to establish a guilty mind according to the crime defined in law will lead to an acquittal; but that will hardly serve the demands of a risk Society for a conviction. Professor Mukherjee was addressing the criminalisation of the seafarer but the principles to be applied in civil aviation are the same; his assertion that prima facie a State enjoys its sovereign prerogative to enact a law that criminalises an aviation accident, or, in the context of his work, an accidental oil spill, effectively criminalises the professional, whatever the circumstances of their risk management.

The imposition of legal standards on human behaviour is somewhat vague in the normative ethics of Society, but Smith and Hogan articulated it perfectly:[12]

> When a citizen is heard urging that, "There ought to be a law against it…," he is expressing his personal conviction that some variety of act is so harmful to Society that it ought to be discouraged by being made the subject of criminal proceedings.

The Courts at common law and the lawmakers in statutory provisions have wrestled with the difficulty of meeting the social imperative with principles of criminal accountability, though, so that the process of turning civil liability in negligence into a crime has trespassed into miscarriages of justice, and the noble aspiration of providing a well-defined, measurable structure for a safety culture, has taken second-place to the demand for punitive Judgments that meet the political demands of Society. Case studies reveal that the process of leaving safety culture to the findings and recommendations in accident reports allows criminal justice systems the opportunity to criminalise the human element, which takes the findings of apparently independent investigations, and weaponises them to support allegations of misconduct or gross negligence for use in the criminal justice system, despite the good intentions of the Chicago Convention. If that does not lead to a criminal conviction, then the award of punitive damages in a criminal case has emerged as another mischief of criminalisation.

The downstream consequence has led to a phenomenon which has been identified as hindsight bias,[13] in which the criminal justice system criminalise professional standards and behaviour in the human element in the aftermath of an accident, on the basis that they were inadequate and now should be punished. The resultant fear of exposure to such threats naturally inhibits the co-operation of individuals in the process of enhancing aviation safety, which is central to the policy of Just Culture.

When analysing the cases studies as we go along, we can draw conclusions that enable us to answer the two questions arising out of Gold's assertion:

- Have aviation professionals undergone a process of criminalisation?
- Has this been an emerging problem for some time?

It is important in this work to ask whether this goes beyond national jurisdictions, and consider how criminalisation has established a transnational basis, in order to satisfy the demands of the risk Society that *there ought to be a law against it.*

In this situation, a major hurdle must be overcome first, because what defines a crime in one State may be inconsistent with the equivalent definition in another. The downstream consequence is that transnational laws may get bogged down in arguments about whether some act or omission should even be considered a crime, particularly if it had not been considered one previously.[14] In their 2021 paper, Aaronson and Shaffer argued that criminalisation research should examine how international and domestic practices of criminal lawmaking interact with one another.[15] Naturally, this is a pivotal question in civil aviation and is certainly important in addressing how the Chicago Convention impacted upon domestic laws.

The relationship between international law and domestic law is the foundation upon which civil aviation law must stand. The essential feature is defined around the High Contracting Parties having signed what amounts to a contract between them. A State which fails to implement the law in its domestic statute book will therefore be exposed to liability to the other States by failing to carry out its duty and implement it. It is a matter between them, and they cannot blame the professional whom they accused of having violated the Convention if it has not been enacted into domestic law. The Convention can only be enforced between those States Parties whose democratic organs have embodied the Convention in domestic law. So, for example, Article 3 of the Chicago Convention provides:

> Each contracting State shall establish all necessary provisions in its national laws or regulations to make such compliance mandatory for any civil aircraft registered in that State or operated by an operator who has his principal place of business or permanent residence in that State. Each contracting State shall make any violation of such applicable laws or regulations punishable by severe penalties and shall submit the case to its competent authorities in accordance with its laws or regulations.

Chapter XVIII of the Convention specifically defines the procedures for disputes and defaults, which emphatically addresses States Parties only. Therefore, legal individuals (people and incorporated bodies including companies) do not acquire any personal rights or obligations under the Convention itself and generally cannot be liable for breaches of it rather than domestic law implemented in accordance with the Convention, unless the State's constitution legislates for that. But the State can also make exceptions to the rule, in accordance with its sovereign rights to legislate for its own people. In a maritime case in the United States Court of Appeal,[16] the Bowater Steamship Company, registered in the UK, sought an injunction to restrain unlawful picketing of their vessel *Gladys Bowater* while alongside in the United States. In support of their claim, they cited a 145-year-old commercial treaty to regulate maritime commerce between the United States and Great Britain, which had been signed and ratified by both States Parties in 1815.[17] Article 1 provided inter alia:

> The Inhabitants of the two Countries respectively shall have liberty freely and securely to come with their ships and cargoes to all such places, Ports and Rivers in the Territories aforesaid to which other Foreigners are permitted to come, to enter into the same, and to remain and reside in any parts of the said Territories respectively, …. and generally the Merchants and Traders of each Nation respectively shall enjoy the most complete protection and security for their Commerce but subject always to the Laws and Statutes of the two countries respectively.

The Court observed that, under the American Constitution, and in accordance with Federal law, when the wording of a treaty is sufficiently explicit to permit its application without additional implementing statutes, and the States Parties intended it to be self-executing, all of which the Court found in this treaty, it can be enforced without domestic legislation. The treaty explicitly conferred the defined rights and obligations not only upon the two States Parties but also upon the inhabitants of the two countries. In this case, the Court held that the United States had to fulfil its treaty obligation of protecting the guaranteed freedoms in two ways: firstly, by not violating its treaty obligations and, secondly, by empowering the Courts to take whatever steps would be necessary to protect and maintain the rights and obligations of individuals, in this case to grant injunctive relief to prevent the Defendants from acting contrary to the treaty.

Article 3 of the Chicago Convention obediently follows the general principle that the Contracting States will enforce the provisions of the Convention by implementing the appropriate domestic legislation as far as their own aircraft are concerned and set the penalties for infringements, where these are applicable. The question of the extent to which the State enacts the Convention provisions in its domestic law may be subject to the legislature's mischievous interpretation of the Convention, in order to circumvent an obligation which may be inconvenient to its purposes. Once again, it is a matter of State sovereignty: a vital feature in the law of Conventions is that the State can implement in its law a higher standard than that set out in the Convention, or the same standard; they cannot, however, implement a lower standard than that set out in the Convention. Consequently, once the domestic law is in force, the prime responsibility for compliance with that law lies with the individual identified in the statute. The pilot in command will be accountable as the owner's representative but will also owe personal responsibility for those laws which the State of registration of the aircraft make them accountable, wherever the aircraft may be, as well as the domestic regime if within their sovereign jurisdiction. The question arises, however, as to which is the appropriate jurisdiction in which to bring criminal proceedings, which is known as the choice of Forum.

State Prosecutors have addressed the Forum issue in deciding whether to bring criminal proceedings in many cases in recent years that are not solely confined to their sovereign jurisdiction. The issue of Forum must be considered during the investigation process, particularly in the matter of arrest, search and seizure and the gathering of evidence, which may fall outside their legal jurisdiction. Naturally, co-operation between States is important, but it is the Prosecutor's responsibility to decide how to proceed in the matter of Forum, and in which jurisdiction to prosecute the offender. The factors to be considered, naturally, depend on the circumstances of the case and Prosecutors must consider how their arrest and detention, and the admissibility of evidence, will meet the demands of the criminal process of the Court in the jurisdiction in question. There are several ways by which a State can claim to have Forum:

- By explicit reference to the jurisdictional reach of the offences created in the Statute;
- By the offence was allegedly committed within their sovereign territory;
- By active personality of the offender; in other words, they are a national of the State concerned;
- By passive personality; in circumstances in which the State has legislated powers to prosecute an offender who has allegedly committed the crime in question against one of its nationals;
- By universal jurisdiction, in which the State has acceded to international laws enabling it to prosecute the offender regardless of their nationality or where the offence was committed.

The jurisdictional issues, therefore, can be truly transnational.

The Prosecutor must make the decision on Forum by taking into account a number of factors, including:

- The location of harm and in what jurisdiction the parties, including the injured parties, were at the time of the offence;

- The interests of the victims and witnesses;
- Whether the Defendant hass connections to the jurisdiction in question;
- Whether the case involves alleged conduct that resulted in very serious harm to individuals in another jurisdiction;
- Whether other suspects or co-defendants are located in the same or overseas jurisdictions and the stage which proceedings overseas have reached;
- Whether multiple prosecutions will be heard in this other jurisdictions;
- Whether any linked prosecutions have started in other jurisdictions and the stages they have reached;
- Whether extradition would make it possible for all prosecutions to take place in one jurisdiction;
- In what jurisdiction the vast majority of evidence is located, whether it can be accessed and whether it can be used in accordance with the jurisdiction's domestic proceedings;
- Where and how investigations may be most effectively pursued;
- Whether there are strong public interest arguments favouring a domestic prosecution;
- Whether the prosecution would suffer a delay in this Forum; and
- Whether this Forum has the legal framework for the management of sensitive information.

Having reached a decision on Forum, the decision to prosecute must necessarily follow the procedure outlined in Chapter 1, which must turn heavily on the evidence. Importantly, however, the Chicago Convention[18] demands that State Prosecutors should take steps to gather evidence to support their own investigations upon which to bring criminal proceedings, rather than attempt to use that obtained by the accident investigation team. This is underlined by a Eurocontrol document published in 2018 which set out a model policy providing directions on conducting a criminal investigation and prosecution of potential criminal offences resulting from aviation incidents, with the objective that *both the safety investigation and judicial process can progress in parallel without either party acting in a manner prejudicial to the interests of the other.*[19] In particular, the guidance emphasises that, where possible under national criminal law, the policy foresees that *no prosecution be brought against individuals for actions, omissions or decisions which reflect the conduct of a reasonable person under the same circumstances, even when those actions, omissions or decisions may have led to an unpremeditated or inadvertent infringement of the law.*

It then qualifies its advice with the proviso that nothing in the policy should prevent criminal prosecutions in the event of intentional wrongdoing or gross negligence, adding in a footnote that *there does not seem to be a commonly agreed definition of gross negligence in Europe.* We are therefore left with the problem raised in Chapter 1, as to whether the facts present an obviously present risk which is clear and unambiguous, not one which might only become apparent after further investigation.

Case study: Tuninter Flight 1153

In this case, on 6 August 2005, a Tuninter Airlines ATR 72, on an international flight from Bari International Airport in Italy, to Djerba-Zarzis Airport in Tunisia, crashed into the sea about 23 nautical miles from the city of Palermo in Sicily. 16 of the 39 people on board died.

The pilot in command was Chafik Al Gharbi, 45 years old, a skilled and experienced pilot with a total of 7,182 flight hours. The co-pilot, 28-year-old Ali Kebaier Al-Aswad, had logged 2,431 flight hours. Both pilots were well-experienced in flying the ATR 72, having accrued 5,582 hours and 2,130 hours, respectively.

The actual aircraft had its fuel quantity indicator (FQI) replaced the night before the flight, but technicians inadvertently installed an FQI designed for the ATR 42, a similar but smaller aircraft with design specification differences for the FQI. As a result, the readings for the fuel tank levels were false, but the problem had not been noticed by the ground crews and the flight engineer, who loaded the aircraft with an inadequate amount of fuel for the flight.

As a result, during the flight, both engines cut out while they were over the Mediterranean. The starboard engine failed first, failing at 23,000 feet, and the port engine then failed at 21,900 feet. The engineer was not a part of the flight crew, but had been called to the flight deck by the pilot after both engines failed (travelling as a passenger, he would die in the crash, though). Captain Chafik requested an emergency landing in Palermo, but the air traffic controller had a poor grasp of English, and failed to assign the distressed flight its own radio frequency on which to communicate, giving the pilots incomplete and/or useless information about their position. These cockpit recordings were omitted from the official investigation report.

Meanwhile, the flight crew repeatedly tried to restart the engines without realising that the fuel had been exhausted. The aircraft glided for 16 minutes, but was unable to reach the runway and they were forced to ditch into the sea, 23 nautical miles northeast of Palermo International Airport. The aircraft broke into three sections upon impact, but the central fuselage remained afloat for some time, enabling rescue vessels from Palermo to arrive in time to save life. Incredibly, only one of the four crew members died, along with 15 out of the 35 passengers.

All of the passengers were Italian, while the crew and the engineer were Tunisian. Autopsies indicated that many of the dead succumbed to the impact, while eight who received injuries during the impact were unable to escape from the aircraft due to their injuries and drowned.

A great deal of the evidence was revealed in the accident report.[20] The crucial evidence concerning the incorrect FQI was found to have originated on the day before the incident, when Captain Chafik (also the pilot in command on that flight) became aware that the FQI on the cockpit display was not working correctly and reported the problem, which was when the FQI was replaced with the incorrect model. The correct FQI was not found because its part number had been entered into the database in a different format that was searched for, and the inventory database mistakenly indicated that the ATR 42 part could be used on both models of aircraft.

Notwithstanding the fitting of the incorrect FQI, suspicions must have been aroused by the quantity of fuel actually loaded in Tunis, but neither the ground crew, the operator nor the flight engineer noticed the difference between the amount of fuel loaded and the change in reading on the FQI. The Pilot in Command had previously raised an anomaly which would become very relevant; when departing from Tunis on the Tunis-Bari route taken before Flight 1153, he noticed by reading his cockpit display that the aircraft's fuel level inexplicably seemed to have increased overnight, but accepted the explanation that there must have been a paperwork error, over which he had no control, but he wrongfully departed without sight of the correct slip.

The accident report also drew a finding which would have implications for the conduct of the flight crew. Routine procedure on the failure of an engine requires that the crew feather the propellers in order to reduce the drag on the blades so that the aircraft can glide a farther distance; but the crew did not feather them, because they were attempting to restart the engines. Furthermore, as a result of their efforts to restart the engines, the crew did not glide the aircraft at the optimum speed to extend the gliding distance.

Captain Chafik was arrested, charged and indicted in a Palermo Court for manslaughter, failing to ensure the safety of the flight and its passengers. Prosecutors had to accept the evidence of the faulty FQI but argued that, after the engines failed, he succumbed to panic, started praying, and failed to follow emergency procedures; they adduced expert opinion that, had he followed procedures, he could have reached the runway at Falcone–Borsellino Airport. In March 2009, he was convicted and sentenced to ten years imprisonment. Six others, including the co-pilot as well as the chief operating officer of Tuninter Airlines, were sentenced to between eight and ten years. The International Federation of Air Line Pilots' Associations protested the flight crews' criminal sentences, alleging that the investigation process was flawed, ignoring mistakes by Italian air traffic controllers who, in the event, were not prosecuted, and there was insufficient evidence against the Captain of deliberate recklessness or gross negligence; as a result, the conviction and sentence resulted from a miscarriage of justice that defines the mischief of criminalisation. The insurmountable difficulty with the complaint of third parties is that sovereign equality of its Member States as guaranteed in Article 2 (1) of the United Nations Charter, which provides immunity from other States' own jurisdiction and their freedom from other States' intervention on their territory.

The system of Roman law practised in Italy gave the Defendants an automatic right of appeal and, in April 2012, the Court reduced the sentences of seven of the Tunisian airline personnel charged. Following their second appeal to the court, Captain Chafik received six years and eight months imprisonment, while the others receiving reduced sentences up to six years. While the air traffic controller in the Tuninter case was not prosecuted, the spotlight of criminalisation was turned on them, as well, which takes us to the criminalisation of air traffic controllers in a very near miss at Amsterdam.

Case study: Amsterdam 1998

On 10 December 1998, an incident occurred at Schiphol (Amsterdam) Airport in which a Delta Air Lines Boeing 767 aborted its take-off roll when the pilots observed a towed Boeing 747 crossing the runway in front of them. At the time of the incident, low visibility procedures were in place. After unclear radio transmissions with the tug driver, an assistant controller had passed her interpretation of the tow's position to the trainee controller responsible for the runway. The assistant controller did not have a display that could show ground-radar images but the trainee controller did and took the position of the tow at the edge of the runway to mean that the crossing had been completed. Buttons on a newly added panel in the tower for controlling lighted stop-bars at runway intersections proved ambiguous, but at the time all looked in order, and he cleared the other jet for take-off. At this time, the coach of the trainee controller was performing supervision duties in the tower.

The Non-Judicial Investigations were carried out by the Incident Investigation Department of Air Traffic Control the Netherlands (LVNL), and a report was published on 4 March 1999. It was some time later that another report was conducted by the Dutch Transport Safety Board (DTSB), which published their report in January 2001. Both investigations reached similar conclusions. In particular, the LVNL report stated on the following contributing factors

- Uncertainty about the operation of buttons on the panel in the control tower and ambiguity with the labelling of the buttons;
- The screen on the Assistant Controller's computer did not show ground radar images;
- The Coach of the Trainee Controller simultaneously had to perform Supervisor duties in the Tower.

The DTSB report identified the following causal factors:

- Low visibility weather conditions which prevented Air Traffic Control from visually identifying vehicles on the ground;
- Inadequate information during the radio communications between the tow-combination and the control tower;
- Misinterpretation of position and movement of the tow;
- Take-off clearance without positive confirmation that the runway was unobstructed;
- Insufficient teamwork and supervision.

THE CRIMINAL PROSECUTION

The Netherlands follows the system of Roman law which we have met in Chapter 1. Two years after the incident, the aviation prosecutor decided to formally charge the coach/supervisor, the trainee and the assistant air traffic controller with the provision of air traffic control in a dangerous manner, or in a manner that could be dangerous, to persons or property.

The outcome in the first criminal hearing was concluded when the Judge sitting alone ruled that the assistant controller was not guilty, but that both the trainee and the coach/supervisor were. They were sentenced to a fine equivalent to $450 or 20 days in prison.

The trainee and the coach/supervisor proceeded to appeal the decision, and the Prosecutor appealed against the assistant controller's acquittal. More than a year later, in September 2002, the appeal was duly heard, during which the three Judges, Prosecutor and the Defence teams had a site visit of the airport's control tower, to get a first-hand look at the work place where the incident occurred. The Court recognised that the facilities in the control tower for the prevention of such incidents were inadequate, as evidenced by the improvements implemented after the incident.

Hearing the appeal, the Court observed that none of the Defendants had a criminal record and that there were no indications that in exercising their responsible functions they had ever failed before. It found all three accused guilty of the charge on the indictment but did not impose any sentence. This reflected a distinction under Dutch law, reducing the criminality of the offence, thus dispensing with the Guilty mind by omitting any reference to a deliberate act, wilful misconduct or gross negligence. In this way, the Court held that there had been an infringement of the law, finding Guilt in the sense that blame was deemed to be present but the essential criminal element was not proved. In simple terms, they were strictly liable, but could not offer a due diligence defence[21] due to the evidence of the *actus reus*, as they were in the control tower and could have taken preventive steps to avoid the commission of the offence. That being said, as no offence had been committed and no punishment levied, there were no grounds for an appeal.

The practical effect has been discussed with pinpoint accuracy by Bert Ruitenberg in his commentary on this case,[22] identifying the asymmetric development between the air traffic controller's duty of care and State prosecutions demanding strict liability. The result instils in their mind the risk that if anything which they do or omit to do in the course of their work, often in times of great stress, subsequently is deemed to be negligent, if not downright dangerous, *they may face criminal prosecution.* As Ruitenberg concluded on the policy of prosecuting air traffic controllers, *A lot of work remains to be done.*[23]

We can compare and contrast this study with a more recent crash, in China in 2016.

Case study: China Eastern flight MU5643

On Tuesday 11 October 2016, a China Eastern Airbus A320-200, on flight MU-5643 was cleared for take-off at Shanghai Hongqiao airport for a domestic flight to Tianjin with 147 passengers. As it was accelerating down the runway, the flight crew saw another Airbus crossing the runway edge and determined that they would not be able to stop in time to avoid a collision. In a rapid crisis assessment, they decided to outclimb the A330, so went to full power and continued their take-off run, rotating and managing to climb the A320 over the A330. The A320 continued its flight to make a safe landing at Tianjin.

The Civil Aviation Administration of China[24] (CAAC) rapidly released a preliminary report, finding that the A320 was in its take-off run when the A330 crossing the runway appeared in front of them. The CAAC very fairly stated that the pilot in command of the A320 assessed that it was safer to continue, concluding with the suspicion that the serious incident was caused by a tower controller's misdirected instruction. The CAAC analysed the evidence that the controller at the west tower of Hongqiao's Airport issued take-off clearance to the A320 in violation of the Basic Flight Regulations of the People's Republic of China and the East China Air Traffic Control's Operations Manual, in particular he did not carry out continuous effective visual observation. He then issued a runway crossing clearance to the A330; then after it crossed the hold short line, the controller instructed the A330 three times to stop, but the crew did not listen or, at least, did not respond; neither did they carry out any visual check to ensure that the runway was clear prior to crossing the hold short line.

The CAAC concluded that the main cause of the serious incident was negligence by the tower controller, but the incident suffered from the contributory negligence of the flight crew of the A330, who did not cross check their instructions, did not check the status of the runway before crossing and did not maintain radio contact with the tower.

On 21 October, the CAAC announced that the tower controller and the supervisor had their licenses revoked for life, but affirmed that the pilot in command of the A320 would receive a first class merit and an appropriate award. Thanks to his actions, what would have been China's worst aviation disaster was averted.

This is not quite the end of the case study. This incident was the subject of a topic of the 2017 Safety Forum, co-hosted by Flight Safety Foundation (FSF), Eurocontrol and European Regions Airline Association (ERA) in Brussels in that June, on the topic of Preventing Runway Collisions, and a presentation made by Captain Pang Ying Qun. In mitigation of the air traffic controller's misconduct, he commented that video evidence suggested the controller was fatigued, but the physical position of the tower also restricted visibility.[25] It highlights the point that negligence is seldom a black and white picture; the air traffic controller and the pilot of the A330 must share the responsibility in the civil tort of negligence, and damages in compensation would be apportioned accordingly. That is not applicable in a crime, though; indeed, how this would work when applying negligence to criminal accountability is not so easy to address, but may be considered in two distinct scenarios:

1. Where such evidence reduces the impact of the criminal elements defined in the offence to the point below the threshold for a conviction;
2. In cases of strict liability, where there is no structure to apportion blameworthiness and harm. Indeed, strict liability may oblige the Court to impute blame to blameless conduct.[26]

The question of fatigue and overwork has been found to be an occupational hazard that can give rise to accountability, and especially in cases of strict breaches of Regulations which involve an unwarranted attribution of blame in a case that can only offer the Defendant mitigation, without a defence in law.

But blame can be difficult to apportion.

Case Study: The Überlingen Crash

This casualty occurred at the night of 1 July 2002 when a Tupolev passenger airliner of Bashkirian Airlines collided with a DHL Boeing 757 cargo aircraft over the towns of Überlingen and Owingen in southern Germany. All 71 people on board the two aircraft were killed.

On the night in question, only one air traffic controller, Peter Nielsen, was in the air traffic control centre in Zurich, covering the airspace through which the aircraft were flying; the other duty controller was resting in another room for the night, contrary to Regulations but a common practice for years, and tolerated by the management of the employing contractor, Skyguide. The ground-based optical collision warning system, which would have alerted the controller to imminent collisions early, had been switched off for maintenance, although Nielsen was unaware of this. There still was the audible Short Term Conflict Alert (STCA) warning system, which released a warning addressed to a workstation 32 seconds before the collision, although this warning was not heard by anyone present at that time, and no system error could be found in a subsequent technical audit. It was academic, though; he could not have resolved the danger in time.

In fact Nielsen was working two workstations at the same time and did not identify the danger in time, so did not take steps to keep the aircraft at a safe distance from each other. Only when he realised the danger at the last minute did he contact the Russian aircraft, Flight 1937, instructing the pilot to descend by 1,000 feet to avoid collision with crossing traffic (DHL Flight 611). Seconds after the Russian crew initiated the descent, however, their traffic collision avoidance system (TCAS) instructed them to climb, while at about the same time the TCAS alarm on Flight 611 instructed the pilots of that aircraft to descend. Had both aircraft followed those automated instructions, the collision would not have occurred.

Flight 611's pilots followed the TCAS alarm and initiated a descent but could not immediately communicate with the air traffic controller because he was dealing with 2937. The Tupolev flight crew had a dilemma; whether to follow the instructions of the air traffic controller, in accordance with Regulations, or to follow the TCAS instruction to climb. They decided to disregard the TCAS instruction and comply with the air traffic control instruction. Nielsen was unaware of the TCAS alerts, and repeated his instruction to Flight 2937 to descend, giving the Tupolev crew incorrect information as to the position of the DHL aircraft. The aircraft collided at almost a right angle at an altitude of 34,890 feet, when the Tupolev exploded and broke into several pieces, scattering wreckage over a wide area, and the DHL Boeing crashed into woods from an uncontrolled descent.

An investigation was carried out by multiple agencies owing jurisdictions in the crash: Switzerland, which had management jurisdiction over the airspace; Russia, where the Tupolev was registered; and Bahrain where the DHL aircraft was registered. They all concluded that miscommunication was a major factor which caused this crash. However, they disagreed on who was to blame, or how it should be apportioned – matters which should not have been in the remit of accident reports in any case. Bahrain found that the private air traffic control provider which employed

Nielsen should have been responsible, as he had been overworked. Switzerland and Russia disagreed on whether the Russian pilots were to blame, as they had the dilemma of whether to follow the TCAS alarm on the air traffic control instruction; and, in the end, they made the wrong call.

In 2004, the German Federal Bureau of Aircraft Accidents Investigation (BFU)[27] published its determination on the accident, and found systemic causes, notably in a failure of the air traffic control provider to ensure that all the workstations were continuously manned, while they also found systemic failures in conflicting Regulations of national aviation authorities, operational and procedures; and that instructions of the TCAS manufacturer and the operators were not standardised, incomplete and actually contradicted each other.

The emerging world of criminalisation demanded accountability. While the air traffic controller, Peter Nielsen, underwent medical treatment for traumatic shock, a criminal investigation of Skyguide's employees was initiated by Swiss prosecutors, leading to charges against the two air traffic controllers on duty and seven other employees of Skyguide on suspicion of gross negligence manslaughter. By the time when the Trial started in May 2007, the number of Defendants indicted was eight, of whom three Skyguide air traffic control managers and one technical project manager were found Guilty. The two air traffic controllers on duty that night were acquitted of gross negligence manslaughter, for the Court had sympathy with the position that a controller acting alone was common practice at a time when low traffic at night prevailed, such as on the occasion of the crash. The managers were held accountable for tolerating negligent, potentially dangerous working practices over seven years, however, which determined the verdict, rather than finding them accountable for the conduct of the air traffic controllers that night. The Prosecutor called for prison terms up to 15 months, but three of the four managers convicted were given suspended prison terms and the fourth was ordered to pay a fine.

CRIMINALISATION IN SWITZERLAND

The rise in the criminalisation of air accidents has been particularly noticeable in Switzerland in recent years. The Swiss Criminal Code was based on an initial draft in 1893, although the Code did not enter force until 1942, when the provision of substantive law was made federal, rather than a matter for cantonal legislation. Since then, the most recent significant revision took effect in 2007 and introduced the power of the Courts to convert prison sentences of less than one year into fines. Indeed, nearly all prison sentences shorter than one year have since been converted into fines, and conditional sentences to conditional fines. The current Swiss Criminal Code provides:[28]

> 1. Unless the law expressly provides otherwise, a person is only liable to prosecution for a felony or misdemeanour if he commits it wilfully.
> 2. A person commits a felony or misdemeanour wilfully if he carries out the act in the knowledge of what he is doing and in accordance with his will. A person acts wilfully as soon as he regards the realisation of the act as being possible and accepts this.

3. A person commits a felony or misdemeanour through negligence if he fails to consider or disregards the consequences of his conduct due to a culpable lack of care. A lack of care is culpable if the person fails to exercise the care that is incumbent on him in the circumstances and commensurate with his personal capabilities.

Nevertheless, there has been a rise in the number of criminal convictions in civil aviation trials that presents a worrying trend which is common throughout jurisdictions that may have the downstream consequence that co-operation with State Prosecutors investigating an incident might be seen as a threat to their defence. The issue, of course, is that under the Roman system in Switzerland, the powers investigating the incident will also be prosecuting a Defendant at Trial, with the power to submit the contents of accident reports in evidence.

In 2019, Andreas Fankhauser identified a number of such cases[29] in which the criminal accountability of the air traffic controller was put on trial. Let us examine one such case.

Case study: Zurich 2012

A Darwin Airline Saab 2000 was departing from Zurich for Genoa, as a private Sportcruiser was on final approach of a training flight under instruction, with the intention to conduct a touch and go. The air traffic controller, who was unnamed in proceedings, gave the Saab take-off clearance on runway 28, while the light aircraft was on its approach, leading to a serious near-miss. The incorrect assessment of the situation by the air traffic controller was established as a direct cause of the incident, in which the tower instructed the Sportcruiser to turn left onto a left downwind for runway 28, then instructed the aircraft to conduct a right hand orbit. The Sportcruiser turned away at low altitude as the two aircraft converged to a lateral distance of 673 feet. The Saab took off and continued for a safe landing in Genoa, as the Sportcruiser managed the intended touch and go on runway 28 several minutes later.

The Swiss Council for Accident Prevention opened an investigation into the close encounter and, on 24 September, released their final report, concluding that the probable causes of the serious incident were:

- The air traffic controller gave a take-off clearance on runway 28 to a commercial aircraft while a light aircraft was on its approach to runway 16;
- The incorrect assessment of the situation by the air traffic controller was established as a direct causal factor for the serious incident.

The following factors were identified as systemic factors contributing to the serious incident:

- Risk assessment by air traffic control was inadequate. A complex training flight was taking place at a time when the volume of traffic was increasing and at medium complexity.

- Because visual approaches were rarely carried out on runway 16, there were only few values derived from experience on which one could have relied when carrying out ones planning.
- The TCAS installed in the Saab did not provide any information to the crew on the ground about other aircraft, which restricted their situational awareness.

The following factors were identified as neither causal nor contributing but were in the context of the investigation recognised as risk factors:

- Air traffic control instructed the light aircraft which was in landing configuration and close to the ground to perform a tight 360 degree turn to resolve the conflict.
- Misinterpreted motivation to provide service on the part of air traffic control.

The Swiss investigation was complicated by the fact that neither the crew of the Sportcruiser nor the tower controller submitted a report. The crew of the Sportcruiser argued that they were not in a position to assess the risk adequately and therefore did not file a report, while the tower assessed the situation as non-critical. Only the crew of the Saab submitted a report and thus initiated the investigation into the occurrence. The investigation team observed that *just because other parties are involved in an incident does not release one from the obligation to submit a report.*

The air traffic controller was duly indicted for the negligent disruption of public transport within the meaning of Article 237 of the Criminal Code:

> 1. Any person who wilfully obstructs, disrupts or endangers public traffic, in particular traffic on the roads, on water or in the air and as a result knowingly causes danger to the life and limb of other people shall be liable to a custodial sentence not exceeding three years or to a monetary penalty. If the offender thus knowingly endangers the life and limb of a large number of people, a custodial sentence of from one to ten years may be imposed.
> 2. If the person concerned acts through negligence, the penalty is a custodial sentence not exceeding three years or a monetary penalty.

At the hearing in the Bulach District Court in March 2019, the Prosecution and Defence disagreed on whether the collision risk had been real or only hypothetical. Two of the three judges held with the Prosecution, finding that the Defendant had violated the air traffic controller's general duty of care and also the general danger principle in a foreseeable and avoidable manner by issuing the take-off clearance to the Saab, by insufficiently monitoring the final approach of the sports cruiser and by instructing the crew of the Sportcruiser to turn right. He had therefore caused the objective concrete danger to life and limb of the crew of the Sportcruiser in a careless manner in breach of his duty. The Court concluded that the Defendant's actions had therefore met all the elements of the offence of negligent interference with public transport within the meaning of Article 237.[30] The Prosecution sought a custodial sentence, to be suspended pending a probation period and by a majority the controller was fined in accordance with Swiss process.

The Defendant appealed to the Federal Court and, on 19 February 2021, the Higher Court of the Canton of Zurich overturned the verdict and acquitted the air traffic controller, finding that he could not be proved to have acted negligently or to have disturbed public transport.

There was a recurring theme in the cases discussed by Fankhauser, concerning the issue defined in the Statute that the Defendant knowingly caused danger to the life and limb of other people, challenging the question as to whether the danger was actual, or hypothetical, and the decisions led Fankhauser to ask the question, whether the evidence merited a criminal prosecution, particularly given that no one was injured and no damage occurred. He made the point that the risk in all the cases was hypothetical, because the pilots took avoiding action and quickly resolved a *potentially* dangerous situation. There may have been questions of a duty of care but, in the absence of wilful misconduct, the most that could be established was the Defendant's negligent care; without evidence of actual loss or damage, the essential provisions in negligence had not been met. The requirements set in the case of Mrs Donaghue's bottle of ginger beer still hold firm[31]; but this has not apparently halted the trend in the criminalisation of incidents.

Perhaps it is the trend which has fine-tuned the trigger finger of prosecuting authorities, with the result that, today, we see the arrest of aviation professionals as the default response in a casualty.

Case Study: LATAM Perú Flight 2213

On 18 November 2022, an Airbus A320neo operated by LATAM Perú was taking off from Jorge Chávez International Airport in Lima, bound on a domestic flight for Juliaca.

The flight received clearance for take-off and commenced its take-off roll on runway 16 at 15:11. During the take-off roll, however, while the aircraft was accelerating, a number of airport crash tenders and fire engines crossed the runway in front it, having been given control tower clearance for a planned emergency drill. The Pilot in Command did not receive any warning of the incursion, and in a split-second decision tried to avoid one of the fire trucks, while another turned right to attempt to avoid the aircraft, but the aircraft collided with one of the trucks, when the starboard engine separated from the wing and the starboard landing gear collapsed. The aircraft then tipped over to its right side as it moved across the runway on fire, and came to rest 8,200 feet down the runway. The evidence clearly showed a fire engine that had been destroyed and the aircraft resting on its right wing with fire damage to the rear fuselage.

The aircraft involved in the accident was a five-year-old Airbus A320neo that had been delivered to LATAM Chile in November 2017. The aircraft was powered by two Pratt & Whitney PW1127G engines, and neither aircraft nor the engines had suffered any defect. The aircraft was damaged beyond repair and was subsequently written off. It was the first hull loss of an A320neo.

Two firefighters in the truck were killed and a third was injured. All 102 passengers and six crew aboard escaped in the emergency evacuation, with 40 being injured. The aircraft was destroyed in the subsequent fire.

Despite their outstanding crisis response in split second decisions which prevented an appalling humanitarian disaster, the pilots of the aircraft were arrested soon after the accident and held in custody until 21 November. The International Federation of Air Line Pilots' Associations (IFALPA) issued a strongly-worded statement concerning the conduct of the Peruvian government on 22 November,[32] stressing the need to avoid prioritising the criminal function over the accident investigation. They pointed out that only a full investigation conducted in accordance with Annex 13 of the Chicago Convention could establish a neutral platform for making findings and, consequently, recommendations, to avoid similar occurrences in the future, and went so far as to offer expertise from their own investigators if the Peruvian authorities needed the relevant assistance in order to meet Annex 13.

IFALPA emphasised the priority for a neutral accident investigation before taking any precipitate action which could injure the human rights of the flight crew. They were particularly alarmed that the flight crew were placed under arrest immediately after the accident for three days; for individuals at the best of time, arrest and detention for three days in a threatening and uncertain atmosphere would have a detrimental effect, but when put under intense psychological pressure in the immediate wake of a horrifying accident, the position in which they were placed was intolerable. IFALPA expressed the strong position that the flight crew should be provided with immediate medical care and psychological assessment by qualified medical personnel before questioning in the immediate aftermath of any accident. Only then can a decision be made determining whether they are fit to give fair and accurate evidence.

How the conditions encountered in this case could have assisted a fair investigation is beyond the comprehension of professionals in their position, and a focus on Annex 13 which subordinates the criminal investigation to the technical one, in order to focus on the protection of safety information from inappropriate use, particularly ensuring that *a balance is struck between the need for the protection of safety information in order to improve aviation safety, and the need for the proper administration of justice*. The positive safety culture enshrined in Annex 13 provides that exceptions to these rules should only be granted by national laws and regulations when there is evidence that the occurrence was caused by misconduct considered to be wilful or reckless, with knowledge that damage would probably result, or the State authority reviews the facts and determines that the release of the safety information to the enforcement authority is necessary for the proper administration of justice, *and that its release outweighs the adverse domestic and international impact such release may have on the future availability of safety information*. That can hardly have been the case here; the arrest was so early that they could not possibly have formed an opinion on the matter. IFALPA also pointed out the Peruvian State's disregard for Annex 19, setting out the safety management processes which must be applied in meeting Annex 13.

On a broader note, IFALPA warned of the danger associated with early arrest and detention, that the public would draw the conclusion that the flight crew was to blame, in other words, scapegoating them before the technical facts or findings had been drawn, at the very least damaging their professional reputation.

IFALPA concluded with an insistence that the State authority comply with the UN's *positive safety culture principles*, which should be a cautionary tale for all criminal investigators who are tempted to give priority to the criminal process.

We may go a little further, and highlight breaches of the Universal Declaration of Human Rights[33] in this case, which were violated on a number of levels:

Article 2: Everyone is entitled to all the rights and freedoms set forth in this Declaration, without distinction of any kind.... Furthermore, no distinction shall be made on the basis of the political, jurisdictional or international status of the country or territory to which a person belongs.
Article 3: Everyone has the right to life, liberty and security of person.
Article 5: No one shall be subjected to torture or to cruel, inhuman or degrading treatment or punishment.
Article 9: No one shall be subjected to arbitrary arrest, detention or exile.
Article 12: No one shall be subjected to arbitrary interference with his privacy, family, home or correspondence, nor to attacks upon his honour and reputation. Everyone has the right to the protection of the law against such interference or attacks.
Article 13: Everyone has the right to freedom of movement and residence within the borders of each state. Everyone has the right to leave any country, including his own, and to return to his country.

The prosecutors' trigger response also involved the suspension of the air traffic controllers. Evidence subsequently emerged that the firefighters were performing an exercise that had been scheduled to commence at 15.10 and was to last ten minutes, with operations co-ordinated by the Peruvian Corporation of Commercial Airports and Aviation, which was responsible for air traffic management. The Peruvian Aviation Accident Investigation Commission released its final report in October[34] 2023, concluding that that the incident was caused by a lack of coordinated planning, poor communication, and the non-use of standardised phraseology. While highlighting the point that the fire crews mistakenly entered the runway when they were supposed to enter a taxiway, it stated that this happened without explicit authorisation from the tower, during a training exercise. The report exonerated the flight crew in its findings but concluded that *the unauthorized vehicular entry into the active runway resulted from a lack of joint planning, poor coordination, and the failure to use communication and standardized phraseology as prescribed by the International Civil Aviation Organization (ICAO).*

No information has yet come to light concerning any criminal action against the air traffic controllers.

CRIMINALISATION AND CORPORATE ACCOUNTABILITY

We were introduced to the concept of corporate accountability in Chapter 1; now we must examine its application in the rising trend of criminalisation.

Having spent some time on Swiss cases, it is appropriate to mention that the famously democratic electorate of Switzerland rejected a proposal to make businesses liable for human rights or environmental violations. However, the demands for corporate social responsibility will not go away.

Corporate social responsibility (CSR) covers many issues of twenty-first century concerns that Society demands companies must take into account, including working practices, management transparency and the human element, that companies must consider for the benefit of all stakeholders, be they shareholders, employees, passengers or ground staff. CSR goes beyond strict regulation though and strives to meet Society's expectations, demanding that companies take responsibility for all of the activities they perform there or abroad in accordance with internationally recognised CSR standards and guidelines. In January 2020, the Swiss Federal Council adopted a revised Action Plan, taking forward its commitment to responsible business conduct:

> The Confederation is committed to the design framework conditions for CSR, raises awareness and supports Swiss companies, strengthens CSR in developing or transition countries, and promotes transparency in CSR activities. The objective of this commitment is to strengthen in the long term the current position of Swiss companies as responsible and competitive players.[35]

The case studies illustrating the crime of corporate accountability in Chapter 1 chart the evolution of statutory regulation, but the course of that evolution shines an uncomfortable light on an underlying trend. The pattern follows a clear path in the relationship between the operator and its professional employees, which formerly had underpinned the concepts of responsibility and shared liability in criminal proceedings for generations, but which have eroded with the emergence of criminalisation in which a company may be exposed to criminal accountability for the acts or omissions of its employees. It was illustrated only too clearly in the maritime casualty of the *Exxon Valdez* in 1989, when the shipowner hastened to lay the blame on the Master in its efforts to avoid criminal accountability. It is this which defines that departure from the traditional relationship between an operator and its professional employees.

CASE STUDY: FLIGHT 981 AND THE EMERGENCE OF CORPORATE ACCOUNTABILITY

Turkish Airlines Flight 981 was a scheduled flight, operated by Turkish Airlines from Istanbul to London Heathrow, with a stopover in Paris. On 3 March 1974, the McDonnell Douglas DC-10 involved suffered an explosive decompression and crashed into the Ermenonville Forest, shortly after it had left Paris, killing all 346 people on board. The subsequent investigation revealed that one of the aft cargo doors on the aircraft was not properly closed and secured. As a result, after take-off, part of the door broke off and caused an explosion in the rear of the aircraft, which damaged cables essential to control flight and led to catastrophic loss.

It was not the first time that such an incident happened on a DC-10; in June 1972, a similar defect was suffered by American Airlines Flight 96, causing an explosive decompression in the aft cargo hold, but the pilot in command had managed to land the aircraft safely. On the ground, it was discovered that the rear cargo door had opened in flight, which caused damage to the fuselage, but not the explosive damage in the case of Turkish Flight 981. Both aircraft experienced uncontrolled explosive decompression when the cargo door latches failed, although the different configuration between the two led to catastrophic consequences for the Turkish Airlines flight. Following the loss of the cargo door, a rapid decompression caused the last two rows of seats to be sucked through a large hole in the plane, severing the navigation cables. The aircraft remained in the air for another 70 seconds while the pilots struggled unsuccessfully to regain control. The pilot in command did what he could to mitigate the descent but the aircraft hit the tree tops at about 434 knots, when the wings broke up, spilling and igniting fuel, exploding and killing the remaining people onboard the plane.[36]

Two serious issues arose addressing liability:

- McDonnell-Douglas had previously known about the flaws with the door design and an NTSB directive after a similar incident had not been implemented. The case for corporate accountability, however, originated in 1970; McDonnell Douglas were conducting on-ground pressurisation tests of the first DC-10 hull when the cargo door exploded open, causing a blast which collapsed part of the cabin floor and destroyed many control cables. It meant that the defective design of the locking mechanism of the cargo door made it possible for the vent door to be apparently closed and the cargo door apparently locked when in fact the latches were not fully closed and the lock pins were not in place.[37]
- Turkish Airlines had allegedly failed to station an engineer on the ground and had reportedly rushed the training process for the DC-10; moreover, the manufacturer alleged Turkish Airlines was liable for having employed a mechanic whom they claimed had deviated from proper procedure.

The case for corporate accountability widened when facts were discovered regarding the duty of care of the Convair division of General Dynamics, which McDonnell-Douglas had sub-contracted to manufacture the fuselage and doors according to its specifications. Evidence revealed that, after the aircraft was already in production, the head of product engineering at Convair wrote a memo urging his superior to refer his concerns about the DC-10's design defect to McDonnell Douglas, predicting that an aircraft would become a catastrophic loss if the defect were not rectified. The engineer's superior replied, however, that McDonnell Douglas already knew about the risk, and if Convair raised it now, they could face liability themselves, because they had owed a contractual duty to the manufacturer to raise safety risks and had failed to do so after the floors had been built and installed on the aircraft. The downstream consequences would involve enormous expense in retroactively changing the floor design, for which Convair would be liable.

The evidence of the engineer's superior shows that McDonnell Douglas were aware of the defect, whether Convair reported it or not, but turned a blind eye, a phenomenon developed in insurance law in which liability requires evidence of a dishonest state of mind on the part of the person who allegedly turned a blind eye to facts which, by their omission, amounted to a breach of trust. It may not necessarily consist of actual knowledge but may consist of a suspicion combined with a conscious decision not to make inquiries which might result in actual knowledge. The emerging claims arising out of the Turkish Airlines crash arose from the commencement of proceedings by the dependents of the 346 occupants, of whom 23 were Americans and the remainder were nationals of 24 other countries. The Defendants joined were Turkish Airlines as the operator, McDonnell Douglas as the manufacturer, General Dynamics for the involvement of Convair and the United States as the manufacturing Regulator. The huge number of cases and jurisdictions had to be consolidated and, in July 1975, the Judicial Panel on Multidistrict Litigation transferred the cases to the Central District of California and assigned them to Judge Hall, who announced that the Defendants had agreed on a Settlement Formula and were willing to negotiate settlements or conduct trials limited to the issue of compensatory damages.

Settlement in the Turkish Airlines case

Most of the cases were settled by the end of the following year, while those cases still deadlocked were referred to a panel of nine retired California Superior Court judges. Total damages exceeding $62 million were settled, excluding legal costs, all of which naturally had a wider, and more serious downstream consequence, for the risk of shareholder flight and bad publicity that could ruin a company. Most importantly of all, the corporate Defendants submitted a motion that the claim for punitive damages could not be upheld in Californian law, but Judge Hall held, that the California law denying punitive damages in wrongful death cases violated the equal protection clause of the American Constitution and the California state constitution, leaving wide open subsequent claims for punitive damages. An appeal of that decision was filed but, in the end, the parties settled their claims. By 2019, over $62 million had been paid to 1,123 Claimants, and the downstream consequences of civil aviation insurance prompted very close industrial analysis.[38]

The damages were all for compensatory loss, to put the innocent parties back into the position that they were in but for the Defendants' wrongs. It did not address punitive damages, but it was in the jurisdiction of California that the issue of punitive damages would very shortly be addressed.

The right to punitive damages follows statutory provision in the State of California. Civil Code section 3294 provides that a Plaintiff can obtain punitive damages when it is proved by clear and convincing evidence that the Defendant has been guilty of oppression, fraud or malice – in other words, that the Defendant had the *mens rea* and the *actus reus* for criminal accountability. Crucially, it is this which harnesses the concept of punitive damages to criminalisation.

The intention of punitive damages is to punish the Defendant in civil proceedings for wrongdoing arising out of a guilty mind; in other words, a criminal wrong. The effect is that such punishment for proving criminality beyond reasonable doubt is addressed in a civil Court whose procedural rules, however, are predicated on deciding a case on the balance of probabilities, often defined in the United States as 'clear evidence' and will award damages instead of the criminal Court's powers under its procedural rules.

Case study: Grimshaw v Ford

In 1968, Ford began designing a new car that would eventually become known as the Pinto. At a time of economic uncertainty, car manufacturers were exploring the market to respond to the public demand for economy. In downsizing the essential requirements to fit in a car, Ford tested the integrity of the fuel system in rear-end collisions. Crash tests revealed that the Pinto could not meet the proposed regulations required impacts without significant fuel spillage, however, with the risk of appalling consequences for drivers and passengers in a fire. They examined the options and found that equipping the car with a reinforced rear structure, smooth axle, improved bumper and additional crush space at a total cost of $15.30 for each vehicle, bringing it within the tolerances of proposed regulations. That, however, would take the car's sale value over the target of $2,000. Then they considered the age-old formula of risk analysis, by considering the probability of an accident which the engineers feared, against the consequences if it does happen, in terms of damages in compensation. The actuarial advice resolved their quandary: it would cost more to remedy the design defects than the damages in compensation which they would likely have to pay to claimants. Ford therefore produced and sold the Pinto to the public without doing anything to remedy the defects.

In 1972, a Ford Pinto was hit from the rear by a car that was following, causing a serious fire exactly as the engineers had feared. The driver of the Pinto suffered severe burns to her entire body and resulted in her death by congestive heart failure. 13-year-old Richard Grimshaw, a passenger, suffered severe, permanently disfiguring burns to his entire body. Grimshaw underwent numerous skin grafts and extensive surgeries, but still lost portions of the fingers on his left hand and his left ear in the accident. Doctors estimated that Grimshaw would require many more surgeries within the next ten years.

Litigation ensued in the State of California,[39] with a claim not just for compensatory damages but also for punitive damages under California Civil Code 3294, as a quasi-criminal punishment on a finding of oppression, fraud or malice. The main purposes of punitive damages are punishment and deterrence of the conduct of corporations which made risk decisions with a guilty mind, providing consumer protection against defectively designed mass-produced articles. In Grimshaw's case, the Court held,

> We find no statutory impediments to the application of Civil Code section 3294 to a strict products liability case based on design defect.

The Jury awarded the Plaintiffs $127.8 million in damages, of which $125 million was accounted for in punitive damages, and $2,841,000 in compensatory damages to Grimshaw; $665,000 was awarded in compensatory damages to the family of the deceased driver. It was the largest award ever in an American product liability and personal injury case, at the time and, although, the Judge reduced the punitive damages, he subsequently said that it was *still larger than any other punitive damage award in the state by a factor of about five.*

This case will be revisited in Chapter 5, analysing the anatomy of accountability, which takes us from the emergence of corporate accountability to its application in the context of Just Culture.

The issues underpinning the criminalisation phenomenon have a direct focus on the determination of what constitutes negligence, and what constitutes gross negligence, and the implications for the prosecution of offenders far beyond pilots and air traffic controllers, as identified by Nemsick.[40] In turn, such issues will have direct consequences upon potential claims for compensatory damages and, in certain jurisdictions, punitive damages. This leads us to the far more detailed Case study, concerning the Lexington crash.[41]

Case Study: the Lexington Crash

Comair Flight 5191 was a scheduled domestic passenger flight from Lexington, Kentucky, to Atlanta, Georgia, operated on behalf of Delta Connection by Comair,[42] a wholly owned subsidiary of Delta, maintaining its own management and policies, and employing its own pilots.

Runway 22 Blue Grass Airport in Lexington is 7,003 feet long, 150 feet wide and lighted for nighttime use. Runway 26 is half the length, at 3,500 feet long, 150 feet wide and only 75 feet wide, not lighted and thus restricted to daytime use only. In order to take off from runway 22 it is necessary to taxi across the end of runway 26. An airport construction project, begun in 2004, was still underway at the time of the accident, necessitating changes to some of the taxiways and signage.

On 27 August 2006, Comair's Bombardier CRJ-100 was due to take off in darkness at 06.00. There was one air traffic controller in the tower; yet again, evidence of a breach of Regulations; in this case, it violated a 2005 memorandum, which required two controllers during the night shift: one in the tower working clearance, ground, and tower frequencies, and the other either in the tower or working on radar at Indianapolis Center. At the time of the accident, the single controller in the tower was performing both tower and radar duties. After handling several aircraft at the beginning of his shift, there were several hours without aircraft movements. In the 20 minutes leading up to the accident, there were three departures, including Comair 5191, under his control. He duly cleared the flight to take off from runway 22 and to fly runway heading 220. After providing take off clearance, he turned away from the window to perform an administrative task and did not carry out further observation; he was not required to.

Evidence from the cockpit voice recorder confirmed that the aircraft was cleared to take off from Runway 22; but after confirming "Runway two-two," Captain Clay taxied to the short line for Runway 26, rather than Runway 22, and remained there for about 50 seconds. The aircraft then attempted to take off the offending Runway 26, and turned the controls over to First Officer James Polehinke for take-off. The air traffic controller did not see the aircraft taxi to the runway.

Based upon an estimated take-off weight of 49,087 pounds, the manufacturer calculated a minimum speed of 138 knots and a distance of 3,744 feet would have been needed for rotation, with more runway needed to achieve lift-off, which of course was far beyond the length of Runway 26. At a speed approaching 100 knots, Polehinke remarked, "There is no lights" referring to the lack of lighting on Runway 26. "Yeah," confirmed Clay, but there was no evidence that either of them considered aborting the take-off as their speed increased. What the investigators found, was that Clay called for rotation as the aircraft reached the end of the runway but they did not have enough velocity for rotation. The aircraft left the end of the runway and struck a berm, becoming momentarily airborne, before it clipped the airport perimeter fence with its landing gear, and collided with trees, separating the fuselage and cockpit from the tail. The aircraft struck the ground about 1,000 feet from the end of the runway, when most of the people of board were killed instantly and the wreckage was engulfed by fire. 49 of the 50 people on board, including Captain Clay; First Officer Polehinke was the sole survivor, with serious injuries including an amputated leg.

The National Transportation Safety Board drew a long list of conclusions, the most relevant to us being:

- The captain and the first officer believed that the airplane was on runway 22 when they taxied onto runway 26 and initiated the takeoff roll.
- The flight crew recognized that something was wrong with the takeoff beyond the point from which the airplane could be stopped on the remaining available runway.
- Adequate cues existed on the airport surface and available resources were present in the cockpit to allow the flight crew to successfully navigate from the air carrier ramp to the runway 22 threshold.
- The flight crewmembers' nonpertinent conversation during the taxi, which was not in compliance with Federal regulations and company policy, likely contributed to their loss of positional awareness.
- The flight crewmembers failed to recognize that they were initiating a takeoff on the wrong runway because they did not cross-check and confirm the airplane's position on the runway before takeoff and they were likely influenced by confirmation bias.
- Even though the flight crewmembers made some errors during their preflight activities and the taxi to the runway, there was insufficient evidence to determine whether fatigue affected their performance.
- The flight crew's noncompliance with standard operating procedures, including the captain's abbreviated taxi briefing and both pilots' nonpertinent conversation, most likely created an atmosphere in the cockpit that enabled the crew's errors.
- The controller did not notice that the flight crew had stopped the airplane short of the wrong runway because he did not anticipate any problems with the airplane's taxi to the correct runway and thus was paying more attention to his radar responsibilities than his tower responsibilities.
- The controller did not detect the flight crew's attempt to take off on the wrong runway because, instead of monitoring the airplane's departure, he performed a lower-priority administrative task that could have waited until he transferred responsibility for the airplane to the next air traffic control facility.
- The controller was most likely fatigued at the time of the accident, but the extent that fatigue affected his decision not to monitor the airplane's

departure could not be determined in part because his routine practices did not consistently include the monitoring of takeoffs.
- The Federal Aviation Administration's operational policies and procedures at the time of the accident were deficient because they did not promote optimal controller monitoring of aircraft surface operations.

The NTSB determined that *the probable cause of this accident was the flight crewmembers' failure to use available cues and aids to identify the airplane's location on the airport surface during taxi and their failure to cross-check and verify that the airplane was on the correct runway before takeoff. Contributing to the accident were the flight crew's nonpertinent conversation during taxi, which resulted in a loss of positional awareness, and the Federal Aviation Administration's failure to require that all runway crossings be authorized only by specific air traffic control clearances.*[43]

Claims for corporate accountability followed from the dependent relatives of deceased passenger Bryan Keith Woodward, who sued Comair, Delta Airlines, the United States for alleged FAA errors and entities associated with Lexington airport, claiming compensatory damages in negligence and punitive damages in gross negligence. There was little independent evidence save that from the cockpit voice recorder, which the Court ordered at a case management hearing, as *one of the few neutral pieces of evidence available to plaintiffs.*

The issue of corporate accountability arose out of the Plaintiffs' claim for punitive damages. Comair applied to the Court to dismiss the claim for punitive damages against them, on the grounds that it had not authorised or ratified the wrongful actions in not conducting a proper pre-taxi briefing, having non-pertinent conversations during the taxi and their other mistakes were contrary to Comair's training and procedures. The Plaintiffs were equally certain that the evidence demonstrated that Comair was grossly negligent in its oversight of the airline's safety programme and that the crew's recklessness was foreseeable, in the meaning that we discussed in Chapter 1.

The first Court hearing found that the Plaintiffs had *demonstrated a genuine issue of material fact … as to the gross negligence of Comair.* The court also found that there was sufficient evidence of the crew's gross negligence to support a finding of punitive damages, including the crew's failure to perform a proper taxi briefing, its violation of a federal regulation and the Comair Operations Manual by engaging in non-essential conversation while taxiing, missing the cues that they were taxiing to the wrong runway, failing to follow proper procedures in determining that the aircraft was on the correct runway prior to commencing the take-off, and in failing to abort the take-off even after the first officer recognised that they were on an unlit runway.

Comair subsequently applied to the Court for reconsideration of the Court's orders made in 2008 and 2009 regarding the penalty award of punitive damages. On 2 February 2011, Karl Forester, Senior Judge, delivered Judgment on Comair's motion. The Court held that that the airline was not liable for punitive damages, *as later-corrected failed tests were not clear and convincing evidence that there were similar incidents from which Comair should have anticipated the pilots' conduct that caused the crash.*[44]

The Plaintiffs further claimed that Comair should be held vicariously liable for the conduct of the pilots. The Court held that *there is no evidence that Comair authorized the pilots to line up and attempt to take off from the wrong runway or that*

Comair ratified that conduct. To the contrary, the overwhelming evidence is that the Flight 5191 pilots violated Comair training, the procedures in Comair's manuals, sterile cockpit rules, and the required taxi briefing for the first flight of the day the Court continued, that Comair had numerous crew surface movement protocols that were designed to establish and maintain situational awareness from prior to taxi all the way out to the runway. Citing the evidence of a Comair witness, the Judge concluded that *He admitted that the Flight 5191 crew failed in that regard.*

Similarly, the Court rejected the Plaintiffs' claim that the case had presented clear and convincing evidence that Comair should have anticipated the reprehensible conduct of the Flight 5191 crew. *Kentucky law requires evidence of a pattern of similar conduct by the employee, such that the employer should have reasonably expected the conduct to recur.* The Court continued, that *In the present case, there was no evidence that either Captain Clay or First Officer Polehinke had previously committed a similar error or had any history of similar conduct during their years flying for Comair.* The Court continued, that *Plaintiffs have not shown by clear and convincing evidence that there were similar incidents from which Comair should have anticipated the pilots' conduct that caused the crash of Flight 5191. There also is no evidence that Comair authorized or ratified the conduct of the pilots.*

As to the Plaintiffs' claim of Gross Negligence by Comair, the Court noted that much of the evidence actually refuted the Plaintiffs' claim that Comair management demonstrated a reckless disregard for the safety of its passengers.

In conclusion on the point of punitive damages, the Court held *that the Court's original decision regarding the sufficiency of the evidence to support a jury instruction on punitive damages noted that Plaintiffs provided "evidence that the conduct of the pilots on this flight was normal and not extraordinary."*

The Plaintiffs' claims for compensatory damages then followed, but their cases were eventually settled.

It will not be lost on the reader that the NTSB had also found that the FAA's guidance on air traffic control operating procedures was deficient. In this context, the NTSB issued the following recommendations to the FAA in April 2007:

- Work with the National Air Traffic Controllers Association to reduce the potential for controller fatigue by revising controller work-scheduling policies and practices to provide rest periods that are long enough for controllers to obtain sufficient restorative sleep and by modifying shift rotations to minimize disrupted sleep patterns, accumulation of sleep debt, and decreased cognitive performance. (A-07-30)
- Develop a fatigue awareness and countermeasures training program for controllers and for personnel who are involved in the scheduling of controllers for operational duty that will address the incidence of fatigue in the controller workforce, causes of fatigue, effects of fatigue on controller performance and safety, and the importance of using personal strategies to minimize fatigue. This training should be provided in a format that promotes retention, and recurrent training should be provided at regular intervals. (A-07-31)
- Require all air traffic controllers to complete instructor-led initial and recurrent training in resource management skills that will improve controller judgment, vigilance, and safety awareness. (A-07-34)

In addition, the Board issued the following recommendation to the National Air Traffic Controllers Association

- Work with the Federal Aviation Administration to reduce the potential for controller fatigue by revising controller work-scheduling policies and practices to provide rest periods that are long enough for controllers to obtain sufficient restorative sleep and by modifying shift rotations to minimize disrupted sleep patterns, accumulation of sleep debt, and decreased cognitive performance. (A-07-32)

While the author has no wish to digress from the main issue of corporate accountability for gross negligence, there is another matter of contributory negligence raised in the case that we should mention. This did not implicate Comair in further liability, but Polehinke and the estates of Jeffrey Clay and Kelly Heyer issued complaints in the Eastern District of Kentucky alleging that Jeppesen-Sanderson, the navigation chart-makers were negligent in providing incorrect charts to Comair.

Jeppesen-Sanderson was a company incorporated in Delaware and doing business in Kentucky, which notably was engaged in the design, manufacturing, distribution, and sale of aeronautical navigational charts. The Court observed that the charts could be used to obtain runway information such as length and take-off distances as well as runway configurations and taxiway configurations. Evidence followed in the proceedings, that, on the date of the accident, the crew for Flight 5191 had the most recent Jeppesen chart for Lexington, which was dated January 2006. This chart showed the prior configuration of the taxiways, not indicating that Taxiway A north of Runway 26 had been closed, but that Runway 26 had to be crossed in order to reach Runway 22. Jeppesen had received information from the National Flight Data Center on 23 June 2006 containing information regarding the changes at Lexington but due to a computer error, this information was not properly entered into Jeppesen's system and an updated chart was not produced. However, the modified chart, which was eventually published after the error was realised, did not indicate the closure of the original Taxiway A; rather, it indicated what the final configuration would be when construction was completed. This would have established a serious case in evidence, if it were proved that the pilots had actually relied on it; but there was no such evidence. The NTSB did make relevant findings in its report, but these did not refer to any reliance on the charts:

- Because of an ongoing construction project at Blue Grass Airport, the taxiway identifiers represented in the airport chart available to the flight crew were inaccurate, and the information contained in a local notice to airmen about the closure of taxiway A was not made available to the crew via automatic terminal information service broadcast or the flight release paperwork.
- The controller's failure to ensure that the flight crew was aware of the altered taxiway A configuration was likely not a factor in the crew's inability to navigate to the correct runway.
- Because the information in the local notice to airmen (NOTAM) about the altered taxiway A configuration was not needed for the pilots'

wayfinding task, the absence of the local NOTAM from the flight release paperwork was not a factor in this accident.

In the absence of any persuasive evidence, they could not persuade the Court that the Jeppesen chart misled the pilots and was a proximate cause of the crash.

NOTES

1 Gold, E (Professor), 2004, *The Protection of Masters and Seafarers from Criminalization: Emerging Problems for the Shipping Industry*, Centre for Maritime Law, University of Queensland, Brisbane Prepared for the 31st Annual Conference 'Navigating the Sea of Change' The Maritime Law Association of Australia and New Zealand Adelaide, South Australia 29 September–1 October 2004.
2 SOLAS V Reg 34 – 1: The owner, the charterer, the company operating the ship as defined in regulation IX/1, or any other person shall not prevent or restrict the master of the ship from taking or executing any decision which, in the master's professional Judgment, is necessary for safety of life at sea and protection of the marine environment.
3 UK Reg (EU) No. 923/2012 (as amended).
4 Lubbock, B, 1919, *The China Clippers*, Brown, Son & Fergson, Glasgow.
5 Ericson, R and K Carriere, 1994, The Fragmentation of Criminology, in D Nelken (ed), *The Futures of Criminology*, Sage, London, p. 89
6 Hudson, B, 2003, *Justice in the Risk Society*, Sage, London, p. 203ff.
7 Bennun, ME, 1995, Prosecuting Professional Pilots in the United Kingdom after November Oscar: Reflections on the Law and Policy. *Journal of Air Law and Commerce* 61 (1995) 331.
8 *R v William Glen Stewart* [1991] Isleworth Crown Court (unreported).
9 See www.core.ac.uk/download/pdf/147637575.pdf.
10 See www.ntsb.gov/investigations/AccidentReports/Reports/AAR9706.pdf.
11 Mukherjee P, 2006, Criminalisation and Unfair Treatment: The seafarer's Perspective. *Journal of International Maritime Law* 12 (2006) 325–336.
12 Ormerod D, 2008, *Smith & Hogan Criminal Law*, Oxford University Press, Oxford, p. 7.
13 See Lawrenson, A and G Braithwaite, 2018, Regulation or Criminalisation: What Determines Legal Standards of Safety Culture in Commercial Aviation? *Safety Science* 102 (2018) 251–262.
14 Reiner, R, 2016, *Crime, The Mystery of the Common-Sense Concept*, Polity Press, Cambridge.
15 Aaronson, E and G Shaffer, 2021, Defining Crimes in a Global Age: Criminalization as a Transnational Legal Process. *Law & Social Inquiry* 46 (2) (May 2021) 455–486.
16 *Bowater Steamship Co Ltd v Patterson and Others, The Gladys Bowater* [1962] 303 F.2d 369 USCA.
17 A Convention to Regulate the Commerce between the Territories of the United States and of His Britannick Majesty, signed and ratified in 1815, 8 Stat. 228, and extended indefinitely on 6 August 1827, 8 Stat. 361. See Bevans C, 1974, *Treaties and Other International Agreements of the United States of America 1776-1949*, Vol. 12, Department of State Publication 8761, Washington, DC.
18 Annex 13 and 19.
19 Eurocontrol, 2018, *Model for a Policy Regarding Criminal Investigation and Prosecution of Aviation and Railway Incidents and Accidents*. See www.eurocontrol.int.

20 ANSV, Final report: accident involving ATR 72 aircraft registration marks TST-LBB ditching off the coast of Capo Gallo August 6th, 2005. See www.ansv.it/wp-content/uploads/2021/09.
21 See Chapter 1.
22 Ruitenberg B, 2002, *Court Case against Dutch Air Traffic Controllers*, LNVL, Amsterdam.
23 Ruitenberg B, 2002, *Court Case against Dutch Air Traffic Controllers*, LNVL, Amsterdam, p. 5.
24 See www.Caac.Gov.Cn.
25 See www.aerossurance.com/safety-management/runway-incursion-shanghai-2016.
26 Coleman J, 2002, *Markets, Morals and the Law*, Oxford University Press, Oxford.
27 See www.faa.gov/sites/faa.gov/files/2022-11.
28 Title 2.2 Article 12.
29 See www.lexology.com/commentary/aviation/switzerland/proton-legal-llc.
30 See www.en.justculture.ch/_files/ugd/6e660a.
31 See Chapter 1; *Donoghue (or McAlister) v Stevenson* [1932] All ER Rep 1; [1932] AC 562.
32 See www.ifalpa.org/media/3861/22prl07-latam-airlines-flight-2213.
33 Universal Declaration of Human Rights, United Nations General Assembly, Paris 1948.
34 Comisión De Investigación De Accidentes De Aviación Del Perú, Informe Final, Accidente De Aviación N° 008-2022, Ciaa-Accid-008-2022, Airbus A320n Cc-Bhb, Latam Airlines Perú SA.
35 Le Conseil federal, 2020, Position et plan d'action du Conseil fédéral concernant la responsabilité des entreprises à l'égard de la société et de l'environnement.
36 For a translation of the report published by the French Secretary of State for Transport, see the UK Department of Trade Report, available online at https://assets.publishing.service.gov.uk/media.5422.
37 A great deal of research and analysis has been published on this case; see for example https://admiralcloudberg.medium.com.
38 See www.businessinsurance.com/article/20190531.
39 *Grimshaw v. Ford Motor Co.* [1981] 119 Cal.App.3d 757, 810.
40 Nemsick R and S Passeri, 2012, *Criminalizing Aviation: Placing Blame Before Safety*, accessed via http://apps.americanbar.org/litigation/committees/masstorts/articles/winter2012-criminalizing-aviation-blame-safety.html.
41 In Re Air Crash at Lexington, KY, 501 F.Supp.2d 902 (E.D. KY 2007) (denying plaintiff's motion to remand the case to state court. The court later dismissed punitive damages claims of plaintiffs whose claims were governed by the Montreal Convention. See In Re Air Crash at Lexington, KY, 27 August 2006, 2008 WL 1909007 (E.D. KY 25 April 2008) NTSB.
42 Comair ceased trading in September 2012.
43 National Transportation Safety Board, 2007, Attempted Takeoff from Wrong Runway, Comair Flight 5191, Bombardier CL-600-2B19, N431CA, Lexington, Kentucky, 27 August 2006. Aircraft Accident Report NTSB/AAR-07/05. Washington, DC.
44 In re Air Crash at Lexington, Kentucky, Civil Action (Master File) No. 5:06-Cv-316 – Ksf (E.D. Ky. 2 February 2011).

3 Criminalisation and Human Rights

OVERVIEW

What is the effect on the professional's human rights? The pilot-in-command of an aircraft has the final authority as to the disposition of the aircraft while in command. It means that nobody can prevent or restrict the pilot-in-command from taking or executing any decision which, in their professional judgment, is necessary for safety of life and for the aircraft.

By the same token, they are accountable for the risks that accompany their command decisions; that has always been the case, but the difference now is that they are being criminalised for what previously had been activity that, at worst, merited nothing more than a civil claim in negligence. This is the key point which provides the foundation for the theory in the next chapter, involving the concept of criminal accountability for negligence.

In recent years, the prosecution of offenders has extended to other professionals, most notably air traffic controllers, who also find themselves defending criminal proceedings.

In this chapter, we shall consider just what defines and limits the individual's human rights, starting with the very essence of what the Defendant is entitled to expect in a criminal Trial. It is noteworthy that the English legal system paved the way which Convention subsequently followed, in Article 5 (the right to liberty and security), Article 6 (the right to a fair trial) and Article 7 (no punishment without law) of the European Convention on Human Rights. With this in mind, we shall examine whether the process of criminalisation has undermined the professional's human rights as a principle of justice which must be satisfied before other political interests are satisfied.

PROSECUTIONS AND THE PROFESSIONAL'S HUMAN RIGHTS

The Universal Declaration of Human Rights was proclaimed in December 1948 as a common standard of achievements for all peoples and all nations. While criminal prosecutions have been more easily achieved by the implementation of strict liability offences, they present compelling arguments in claims of unfair treatment of Defendants. Long before the UK adopted the European Convention on Human Rights, a Code of Practice evolved which defined the key features of a fair trial. The most recent iteration is found in the UK's Courts and Tribunals Judiciary *Guide to*

DOI: 10.1201/9781003489696-3

Judicial Conduct[1]which emphasises the three imperatives which are essential to a fair trial, demanding from judges:

- Judicial independence
- Impartiality
- Integrity

The Criminal Procedure Rules define the practices which are essential in criminal court procedure in magistrates' courts, the Crown Court and the Court of Appeal. Part I sets out the overriding objective in all cases, that *criminal cases be dealt with justly*. Among other factors, it demands *acquitting the innocent and convicting the guilty*; *dealing with the prosecution and the defence fairly*; *and recognising the rights of a defendant, particularly those under Article 6 of the European Convention on Human Rights*.[2]

Of paramount importance in achieving this is the independence of the Trial Judge; in other words, they must be seen to be independent of the lawmakers in Parliament and the executive of the civil service which is responsible for implementing the decisions of the lawmakers. This principle of the separation of powers remains fundamental to the concept of the English Legal System, and if Judges are not independent of the lawmakers and the executive, then the system of justice in a free State breaks down. The function of the Court is to interpret the Will of Parliament in applying the law and must be independent of all sources of power or influence in society, which includes the media and the social pressure applied by twenty-first-century communications.

Impartiality is essential to the proper discharge of the Judge's duties in good faith. It applies not only to the decision itself but also to the process by which the decision is made and follows an ancient maxim in English law that no person man in permitted to be judge in their own cause, so there cannot be a conflict of interest. It necessarily follows that the Judge must be free from any bias, which naturally emerges in cases in which popular or social pressures run high, which is a natural phenomenon in cases such as air accidents involving allegations of manslaughter. Naturally, the suspicions of bias can only be determined after the Trial, and a 2001[3] case in the Court of Appeal laid down a number of principles which must be satisfied:

- If a Judge is shown to have been influenced by actual bias, their decision must be set aside.
- Where actual bias has not been established, the personal impartiality of the Judge will be presumed.
- The Appeal Court has to decide whether, on an objective appraisal, the material facts give rise to a legitimate fear that the Judge might not have been impartial; if they do, the decision of the Judge must be set aside.
- The material facts are not limited to those which were apparent to the Appellant; they are those which are ascertained upon investigation by the Court.
- An important consideration in making an objective appraisal of the facts is the desirability that the public should remain confident in the administration

> of justice; whether or not there was a legitimate fear that the Judge might not have been impartial will depend on whether a fair-minded and informed observer would conclude that there was a real possibility that the Judge was biased.

It is noteworthy that, throughout these principles, there is not a word about the pressure that might be imposed by Parliament or the executive; the only thing for the Judge to manage is to apply the facts to the law.

The question of the impartiality of a jury is far more difficult to resolve because the deliberations of a jury are sacrosanct. If, at the time when a jury is sworn in, the Defence has grounds to suspect that a juror may be biased through malice, then they can challenge the selection of that juror, but after the jury has been sworn in, the Judge may discharge a juror on discretionary, but really exceptional grounds. Only rarely will jury misconduct be established; such an example was the conviction of Stephen Young. He had been arraigned on an indictment for murder at the Crown Court in Lewes in 1994. He was convicted, but a juror subsequently contacted the defence team that, after the jury had been sent to a hotel overnight to consider their verdict, a number of the jurors allegedly consulted an Ouija board to ask the two deceased victims the obvious question. They duly returned a verdict of Guilty, but when the defence team was informed of the misconduct, Young appealed against his conviction, and the Court of Appeal allowed his appeal and ordered a retrial. Even though ordinarily the Court of Appeal could not inquire into the deliberations of the jury, the use of the Ouija board had taken place in the hotel and was not part of the deliberations in the jury retiring room. The verdict clearly was unsafe, and a retrial was ordered.[4]

The right to a fair trial goes beyond bias, and Procedure Rules explain what is meant by the Court's overriding objective to deal with the case justly. The importance of a fair trial will include provisions under long-standing English rules on natural justice in procedural fairness:

- A person accused of a crime, or at risk of some form of loss, should be given adequate notice about the charges and sufficient time to prepare their case for Trial.
- A person making a decision should declare any personal interest they may have in the proceedings.
- The Defendant is entitled to know all the evidence on all the counts in the indictment against them, and by the same token must be entitled to answer every allegation made against them.
- The Defendant is entitled to challenge the prosecution evidence in cross-examination (and vice versa) before the Jury, which must balance their conclusion on what they hear (or see).

Certainly, these long-standing rights of natural justice under English law formed a convenient foundation for the European Convention on Human Rights, upon which the Human Rights Act 1998 is based. In December 2020, the government announced an independent review to consider how the Human Rights Act is working in practice and whether any change is needed. The UK's departure from the EU undoubtedly influenced the review's key themes:

- the relationship between domestic courts and the European Court of Human Rights;
- the impact of the 1998 Act on the relationship between the judiciary, the executive and the legislature

This has particular significance for the Pilot in Command, because of the decision of the European Court of Human Rights in the cases of Captain Mangouras, which we shall examine in more detail later; suffice to say, here, that the Court held that his rights were made subservient to the political demands of Spanish governance, giving priority over his human rights to *the growing and legitimate concern both in Europe and internationally in relation to environmental offences and the tendency to use criminal law as a means of enforcing the environmental obligations imposed by European and international law.*[5]

Whatever the influence of English law and practice might have been, it is the globalisation of society which has driven the evolution of international law, bringing its normative ethics to conventions that may have the effect of weakening human rights by facilitating criminal convictions, such as with strict liability offences, and the evolution of gross negligence that has grown as an alternative to strict liability.

THE EUROPEAN CONVENTION ON HUMAN RIGHTS AND ITS PROTOCOLS

Commenting on criminalisation in global terms, Aaronson and Shaffer made the point that many areas of law have been driven by the normative ethics of twenty-first-century society, which demands the need for some practical process of criminalisation for some behaviour that a changing society wants to prohibit. The question of how they challenge individual human rights is determined according to society's demands, and that ends up influencing transnational law.[6] The justification is clear: global society understands the need to make Regulations which protect the safety of aviation and, thus, a violation of those regulations cannot go unpunished. But if the offence does not share the characteristics of a crime, that is, *mens rea* need not be proved, then the sanction must be appropriate to that.

It is the normative ethics of the society of the prosecuting State which determines how the individual actors are to be punished for the harm which society attributes to them, notwithstanding their blameworthiness. In so many of the cases studied in this book, though, the punishment visited upon the Professional is disproportionate, evidencing justification for alarm in the emerging criminalisation of the Professional as the individual responsible, even though they may not have the *mens rea* for blame.

The European Convention on Human Rights is an international treaty between the States of the Council of Europe. The UK was one of the States that drafted the Convention and was one of the first States to ratify it in 1951. The Convention came into force in 1953, although the UK did not bring the Human Rights Act into force under English law until 1998. There are 14 Articles under the Convention, the most relevant ones for us providing as follows:

Article 5 is important in that it defines the basic right to liberty, particularly addressing the individual's right to liberty and security before Trial. We shall address this further in the case study of Captain Mangouras, below.

Article 5

Right to liberty and security:

1. Everyone has the right to liberty and security of person. No one shall be deprived of his liberty save in the following cases and in accordance with a procedure prescribed by law:
   a. the lawful detention of a person after conviction by a competent court;
   b. the lawful arrest or detention of a person for non-compliance with the lawful order of a court or in order to secure the fulfilment of any obligation prescribed by law;
   c. the lawful arrest or detention of a person effected for the purpose of bringing him before the competent legal authority on reasonable suspicion of having committed an offence or when it is reasonably considered necessary to prevent his committing an offence or fleeing after having done so…

Articles 6 and 7 of the European Convention clearly owe their lineage to the English rules of natural justice and are enshrined in the Human Rights Act 1998 and defines the right to a fair trial.

Article 6

1. In the determination of his civil rights and obligations or of any criminal charge against him, everyone is entitled to a fair and public hearing within a reasonable time by an independent and impartial tribunal established by law. Judgment shall be pronounced publicly but the press and public may be excluded from all or part of the trial in the interest of morals, public order or national security in a democratic society, where the interests of juveniles or the protection of the private life of the parties so require, or to the extent strictly necessary in the opinion of the court in special circumstances where publicity would prejudice the interests of justice.
2. Everyone charged with a criminal offence shall be presumed innocent until proved guilty according to law.
3. Everyone charged with a criminal offence has the following minimum rights:
   a. to be informed promptly, in a language which he understands and in detail, of the nature and cause of the accusation against him;
   b. to have adequate time and facilities for the preparation of his defence;
   c. to defend himself in person or through legal assistance of his own choosing or, if he has not sufficient means to pay for legal assistance, to be given it free when the interests of justice so require;
   d. to examine or have examined witnesses against him and to obtain the attendance and examination of witnesses on his behalf under the same conditions as witnesses against him;
   e. to have the free assistance of an interpreter if he cannot understand or speak the language used in court.

Article 7

1. No one shall be held guilty of any criminal offence on account of any act or omission which did not constitute a criminal offence under national or international law at the time when it was committed. Nor shall a heavier penalty be imposed than the one that was applicable at the time the criminal offence was committed.
2. This Article shall not prejudice the trial and punishment of any person for any act or omission which, at the time when it was committed, was criminal according to the general principles of law recognised by civilised nations.

THE PROSECUTION PROCESS AND HUMAN RIGHTS

The hurdle to overcome in criminal defence is to convey a compelling argument that the Prosecution is offensive to the Defendant's human rights in the sovereign jurisdiction of the prosecuting State, for this is where the mischief lies. In understanding the mischief, it is essential to analyse how the normative ethics of society have influenced the current law, in which the professionals find themselves subject to what the aviation community perceives to be a growing phenomenon of criminalisation, notably in relation to prosecuting State jurisdiction but, in fact, the mischief is identified as spreading far beyond, and discredits – shames, even – the most revered of judicial systems.

This may be illustrated starkly by a review of the case of Adomako[7] which we discussed in Chapter 1, when considered in the light of the wisdom of Mr Justice Holmes in the *Northern Securities* case,[8] who said *Great cases like hard cases make bad law. For great cases are called great, not by reason of their real importance in shaping the law of the future, but because of some accident of immediate overwhelming interest which appeals to the feelings and distorts the judgment.*

If we read this in the context of this book, this challenges society to convey a compelling decision to make the Pilot in Command or the Air Traffic Controller accountable for wrongs which the risk society perceives it to have suffered – simply because of the fact of their professional responsibility, which satisfies the demand for a scapegoat for the consequences of an event, rather than demanding any characteristics of a crime for the event.

While this explains the rationale of the risk society for the criminalisation of the aviation professional, it does not render the rationale safe, because of the lack of evidence to support the logical conclusion of the inherent argument, that the distribution and control of risks will restore a balance dictated by society's conscience; without such evidence, one is left to draw the conclusion that the balance in favour of the individual's human rights to justice and fairness would simply be eroded.

There is a prima facie persuasive argument in submitting that the solution would be to implement globally a common ethical system based on the rights theory, that is, the theory that the right of one person imposes a duty on another. Despite the UK's late implementation of the 1950 European Convention on Human Rights, compelling arguments might be adduced to support the pre-eminence of the English Legal System as the torch-bearer of justice in human rights. The English system can be defined as being based on Common Law, having evolved from spontaneously observed rules and practices, shaped and formalised by decisions made by judges pronouncing the law in relation to the particular facts before them. Strong evidence underpinning this can be found in the Rules of Natural Justice, which were largely unwritten rules defining the assumptions of justice that can be relied on in the English system. These rules formed the foundation for Article 6 of the 1950 Convention, requiring that, whether the proceedings be civil, compensating the loser, or criminal, punishing the offender, everybody, whether the Accused or the Accuser, is entitled to a fair hearing in which they can defend themselves, in full view of the public eye, within a reasonable time, before an independent and impartial tribunal which has been established and managed by Statute, that is, by the Will of the People.

The English Legal System embraced and enforced the terms of the Convention by the Human Rights Act 1998, although this was not the first codification of the right under English law (that belonged to Magna Carta), so that, taken as a whole, an argument may be made for applying the English criminal process (as opposed to constituent laws) worldwide in order to determine the criminal liability of the individual.

This, however, would defy a major ethical issue: In 1918, President Woodrow Wilson of the United States addressed a joint session of Congress in which he declared the inalienable right of people to choose freely the politics by which they are governed, and to follow the social development as they choose. It necessarily follows that they must be free to shape their own culture, which means making their own laws and, if it so happens, be free to make their own mistakes. That's democracy for you.

In consequence, it would offend the most basic ethical values of the transnational forum to tender a solution which conflicts with the rights of sovereign States to determine their own judicial process; that is, the system which tries the Accused for liability under their chosen system. Moreover, a solution in which one-size-fits-all cannot be shoehorned into the world's legal systems. There are irreconcilable differences between the adversarial system, led by English law, and the inquisitorial system of most European Continental states who would support their system with the compelling authority of Cicero and, later, Imperial Rome.[9] The task of determining which is the preferred system is hardly one which ethically can be imposed upon a sovereign State against its will, and an attempt to do so can lead to seismic problems, ultimately leading to the UK's departure from the EU.

By the same inalienable freedom of choice, there is nothing to stop a sovereign State from exercising the equal freedom of picking and choosing which legal principles and processes it admires in other jurisdictions. While English criminal procedure involves patterns and processes which are dependent upon the adversarial system, principles of English law have been embraced globally, perhaps the most notable example being the re-statement of civil liability in negligence held obiter in a case in which the villain of the piece was a dead snail lurking in a bottle of ginger beer.[10] Paradoxically, the two galaxies of criminal and civil law clash where they meet this very case, which gave birth to the concept of criminal negligence, and fundamental to the golden thread running through this work that the emergent criminalisation phenomenon, at the centre of the primary research question, is fatally flawed by its reliance upon criminal negligence in finding a conviction against the aviation professional that conveniently suits the normative ethics of the people demanding retribution.

State responsibility also demands the application of risk management to the rights and obligations of air operations. The very fact of the physical place where the accident took place brings the aircraft within the jurisdiction of that State and, necessarily, all the State's laws by which it controls its subjects. When addressing the human rights of individuals to go about their lawful business, it would naturally be preferable to have such laws codified by statute, but, in a Common Law system, many such human rights and obligations have been defined not just in Statutes but also in Judgments and, by their evolution, the aviation professional has some guidance for divining their accountability. The place of prosecution, therefore, will play a distinct part in how that accountability is determined, and facts which are established in Court may explain conduct in mitigation, even taking the case beyond an error of judgment into the realms of negligence, but the risk is then exposed that the pilot takes some decision in the heat of the moment which, in the absence of such pressure, they might have avoided; nevertheless, the human element involved in such a crisis situation may present the Prosecution with the clear characteristics of a crime and, in the eyes of their judicial system, a conviction must be followed by punishment.

In the UK, the CAA emphasises the importance of Annex IV to Regulation (EC) No. 216/2008 (Essential Requirements for air operations), restating that the pilot-in-command is responsible for the operation and safety of the aircraft and for the safety of all crew members, passengers and cargo on board, from the time that they assume responsibility for the aircraft and passengers prior to a flight until the passengers are deplaned and escorted out of the operational area of the aerodrome or operating site and they relinquish responsibility for the aircraft at the end of a flight the flight.[11]

The guidance emphasises the importance of the pilot's position, as being central to their management role in maintaining the law of the registered State on board the aircraft, the sovereign territory over which other States have no prima facie right in which to intervene – unless, of course, their laws have been broken and, very possibly, their nationals have been killed. In either case, the pilot may optimistically take some comfort in their human rights of being able to defend themselves and, indeed, they may be acquitted at the eventual trial. But, they would have to go through that accountability process in the first place and, in the meantime, may be held in custody in some foreign jurisdiction whose accommodation for Defendants awaiting trial is inconsistent with what a pilot may feel entitled to expect. And then, they may still be found Guilty at the trial. No criminal defence can ever be certain.

To summarise, the aviation professional's accountability to the State is born of necessity, in order that the State can maintain management control, either because the aircraft was registered there or the accident took place there – or that the lives of nationals of the State were lost in the accident. It is, perhaps, ominous that the UK Civil Aviation Authority's *Code of Practice for the Investigations and Enforcement Team*[12] makes no mention of human rights whatsoever.

The phrase itself embraces the basic rights and freedoms that belong to every person in the world, from birth until death, and goes far beyond the ambit of the criminal law. In 2013, the European Court of Human Rights ruled on an appeal by Nadia Eweida, a member of check-in staff for British Airways (BA), who was punished for wearing a small silver crucifix on a chain around her neck, as a sign of her commitment to her Christian faith. Her employer sent her home and suspended her without pay, on the grounds that her cross violated company uniform policy. Her complaint through the UK courts was rejected, but the European court ruled that the cross had been discreet and could not have detracted from her professional appearance. In these circumstances, there was no realistic evidence that it encroached on the rights of others.[13]

In a far more serious case for aviation professionals, however and, possibly, without even considering it, the European Court process rationalised the trend of criminalisation, to exclude it from the context of Article 6(2) of the European Convention on Human Rights, that everyone charged with a criminal offence shall be presumed innocent until proved guilty according to law. This was in the prosecution of Captain Mangouras in the maritime case of the *Prestige*, which set a chilling precedent. The Judgment shone a light on the double standards in which human rights are defined and then ignored when it suits the State to do so.

Case Study: Captain Mangouras of the *Prestige*

Captain Apostolos Mangouras was Master of the *Prestige* which, in November 2002, was involved in a catastrophic event which led to a spill of 70,000 tonnes of fuel oil into the sea, endangering marine life and, consequently, leading to a criminal investigation by the Spanish authorities. In the meantime, the Spanish Court remanded Captain Mangouras, a Greek national, in custody and set his bail at an incredible 3 million euros. The case was appealed to the European Court of Human Rights on the grounds of an infringement of the European Convention on Human Rights,[14] notably Article 5 which provides that *Everyone has the right to liberty and security of person. No one shall be deprived of his liberty save in the following cases the Article further states that the lawful detention of a person after conviction by a competent court; the lawful arrest or detention of a person for non-compliance with the lawful order of a court; the lawful arrest or detention of a person effected for the purpose of bringing him before the competent legal authority on reasonable suspicion of having committed an offence or when it is reasonably considered necessary to prevent his committing an offence or fleeing after having done so.*

Captain Mangouras did not meet any such category; the European Court confirmed that a remand in custody with such a high bail could only be required if reasons justifying that detention prevailed, and that the authorities had to take as much care in fixing appropriate bail as in deciding whether the Defendant's continued detention was indispensable. But the European Court went on to uphold the political imperative which the Port State held above the human rights of the Master and expressed the opinion that new realities had to be taken into consideration in interpreting the requirements of the Human Rights Convention, namely what it referred to as the growing and legitimate concern both in Europe and internationally in relation to environmental offences and the tendency to use criminal law as a means of enforcing the environmental obligations imposed by European and international law. The European Court held the view that the increasingly high standard which was demanded by individual human rights required greater firmness in assessing breaches of the fundamental values of democratic societies. In consequence it held,

> The Court could not disregard the growing and legitimate concern both in Europe and internationally about offences against the environment. It noted in that connection the States' powers and obligations regarding the prevention of marine pollution and the unanimous determination among States and European and international organisations to identify those responsible, to ensure that they appeared to stand trial and to punish them.[15]

Clearly, the European Court upheld the sovereign jurisdiction of a domestic court to manage its judicial process and declined to take any step that could be interpreted as an intervention, despite the inconsistency of the decision with Article 6 (2) of the European Convention on Human Rights which demands that *Everyone charged with a criminal offence shall be presumed innocent until proven guilty according to law.*

The moral of the story lies in the inescapable truth of the criminalisation phenomenon, that, if passed within the judicial process of a sovereign state, it may not be justice, but it is the law, and it is this tension between justice and fairness which has brought the topic of criminalisation to the top of the agenda. The Court in any democracy, after all, is bound by the constitutional parameters of its powers, which have been accorded to it by the State, that is, by its own People in a democratic system. If the State loses confidence in the international process to serve its own Society, then its own domestic law, rather than its international obligations, will prevail, rather than the other way around. International agreements of human rights are only binding if a Court of sovereign jurisdiction embraces them; that must be the case, because the European Court has said so.

REVISITING THE CASE OF CHAFIK AL GHARBI

In Chapter 2, we discussed the case of Tuninter Flight 1153. Captain Chafik was arrested, charged and indicted in a Palermo Court for manslaughter, failing to ensure the safety of the flight and its passengers. Prosecutors accepted the evidence of the faulty FQI but argued that, after the engines failed, he succumbed to panic, started praying and failed to follow emergency procedures. He was duly convicted and sentenced to ten years imprisonment.

It is worth reminding ourselves of Article 7 of the European Convention on Human Rights:

> No one shall be held guilty of any criminal offence on account of any act or omission which did not constitute a criminal offence under national or international law at the time when it was committed. Nor shall a heavier penalty be imposed than the one that was applicable at the time the criminal offence was committed.

The crime of Manslaughter under Italian law demands evidence of the intention of harm to the deceased, although they did not intend to kill them. International Federation of Air Line Pilots' Associations protested strongly but with calm deliberation that *there was insufficient evidence of deliberate recklessness or gross negligence; as a result, the conviction and sentence resulted from a miscarriage of justice, that defines the mischief of criminalisation.*[16]

IFALPA argued that Captain Chafik and First Officer Ali Lassoued performed their duties well above normal levels, and the Prosecution relied on hindsight and knowledge which the crew on that day were denied, rendering unsafe any allegation of recklessness or gross negligence. In these circumstances, the evidence leads to the conclusion that the Defendants were denied the right to a fair trial under Article 6. The trouble is that no amount of protest by IFALPA can interfere in the legal process of a sovereign State. It was enough that the flight crew received reduced sentences on a second appeal. The Italian air traffic controllers were not prosecuted at all, bringing the fairness of the prosecution into further doubt, raising the question of political motivation for the purpose of meeting pressure on government control.

POLITICAL MOTIVATIONS

Human rights can have severe repercussions on States in the international context that flout them for political purposes. State sovereignty allows them to implement their own rules on aviation within their jurisdiction but that implementation must be conducted and managed within the demands of the Chicago Convention. Article 1 states that the contracting States recognise that every State has complete and exclusive sovereignty over the airspace above its territory. The Convention gives States acting in good faith the right to divert or intercept foreign aircraft for security reasons. The right is limited, however, and should be subject to the Standardised European Rules of the Air (SERA), which provide:

> In case of an aircraft known or believed to be in a state of emergency, including being subjected to unlawful interference, ATS units shall give the aircraft maximum consideration, assistance and priority over other aircraft, as may be necessitated by the circumstances.
>
> Subsequent ATC actions shall be based on the intentions of the pilot, the overall air traffic situation and the real-time dynamics of the contingency.[17]

Clearly the authority and responsibility of the pilot-in-command must be upheld, and under no circumstances subject to threats of violence or false pretences. Which brings us to the case of Roman Protasevich and the Belarusian government.

Case Study: Ryanair Flight 4978

This was a regularly scheduled international passenger flight from Athens International Airport, Greece, to Vilnius Airport, Lithuania, operated by the Polish subsidiary Buzz. On 23 May 2021, it was overflying Belarusian airspace, with no plan to land, when it was diverted by the Belarusian government to Minsk National Airport which instructed a Belarusian air traffic controller to supply the false information of a claim of a Hamas bombing attempt. While detained at the airport, no bomb was found but two of its passengers, opposition activist and journalist Roman Protasevich and his girlfriend Sofia Sapega, were arrested and removed for detention.

The aircraft was allowed to depart after seven hours, reaching Vilnius eight and a half hours behind schedule.

Article 1 of the Chicago Convention provides that *Every state has complete and exclusive sovereignty over airspace above its territory* and Article 6 states *No scheduled international air service may be operated over or into the territory of a contracting State, except with the special permission or other authorization of that State*. However, Article 4 states that *Each contracting State agrees not to use civil aviation for any purpose inconsistent with the aims of this Convention.* This underlines the First Freedom Right which defines the right or privilege, in respect of scheduled international air services, granted by one State to another State or States to fly across its territory without landing.[18]

On 27 May 2021, the International Civil Aviation Organization (ICAO) Council decided to undertake a fact-finding investigation[19] under the Chicago Convention, which established that established that Belarus' actions amounted to the use of civil aviation for a purpose inconsistent with the aims of the Chicago Convention and in contravention of its article 4 on the misuse of civil aviation. The ICAO Assembly subsequently adopted Resolution A411, condemning the actions of the government of Belarus.

As an apparent abuse of the State's rights, the subterfuge and detention was condemned by the EU, the UK, the United States and others. In particular, a State investigation by Poland established that it had been an act of piracy and illegal imprisonment, while Lithuania initiated an investigation with regard to criminal offences against Personal Liberty, Honour and Dignity in Chapter of its Code of Criminal Law. The United States became directly involved because four American nationals were on the aircraft; they went somewhat further, when an investigation by the FBI recommended an indictment under United States law charging four individuals who were government officials of Belarus with conspiracy to commit aircraft piracy, an offence which, on conviction, carries a maximum penalty of life in prison and a mandatory minimum sentence of 20 years.

Taking the issue to another level, the case was referred to the Security Council,[20] on the complaint that the government of Belarus' deliberate diversion of Ryanair Flight 4978 under the false pretext of a bomb threat was an act of unlawful interference that deliberately endangered the safety and security of the flight and the lives of those on board. Council members condemned Belarus' violation of international aviation law, while others warned that the serious precedent set by such action undermined the fundamental trust between pilots and air traffic controllers and jeopardised the integrity of international civil aviation. Speakers also drew attention to the government's repression of civil society and human rights violations, noting that the purpose of the dangerous operation was to arrest opposition journalist Roman Protasevich and his companion, Sofia Sapega.

Speakers made representations from France, Germany, Greece, Albania, the UK, the United States, India, Ireland, Norway, Kenya, Mexico, the United Arab Emirates, Ghana, Gabon, Lithuania, Poland and Latvia (also on behalf of Estonia and the Czech Republic). Brazil's representative voiced concern about indications of Belarus' violations of the Convention for the Suppression of Unlawful Acts against the Safety of Civil Aviation, but suggested that a more appropriate organ would be the Office of the High Commissioner for Human Rights. China's representative was equally circumspect, and advised the Council the matter should be handled pursuant to the purposes and principles of the Charter of the United Nations and international law.

Naturally, Russia officially backed Belarus, and defied the international condemnation, offering the not unexpected allegations of international bias. Ultimately, there was little that could be done, save that the EU and the European Union Aviation Safety Agency issued directives halting European airlines from flying over Belarusian airspace.

In May 2022, a court in Belarus sentenced Sofia Sapega to six years in prison on charges of organising riots, calling for acts of terrorism and insulting Belarusian President Alexander Lukashenko. As a Russian citizen she was subsequently transferred to Vlaidivostock, Russia, to serve her sentence. In February 2023, Protasevich initially pleaded Guilty, but this somewhat encouraged the Prosecutors who got the bit behind their teeth and, in April, they indicted him on new criminal charges, including the charges of inciting social hatred, organising mass riots, and organising and running extremist groups which called for the overthrow of the Lukashenko government. Protasevich then pleaded Not Guilty. Thirteen days later, he changed his plea to Guilty and offered in mitigation that he had been manipulated by others to commit the offences, and it was alleged that he incriminated Sofia Sapega in a confession. On 3 May 2023, Protasevich was sentenced to eight years in prison by the Minsk District Court on charges including plotting a coup; but three weeks later he was apparently pardoned. A month later, it was announced that Sapega had been pardoned. By this time, of course, events in the region had taken an altogether more serious course and the war in Ukraine had been raging out of all proportion to that which had been expected when the Russians invaded in February 2022.

In the UK, the fairness of the trial proceedings was found to be inconsistent with Article 6 of the European Convention on Human Rights. This arose in part from allegations that Protasevich had been pardoned following a forced confession and testimony incriminating Sofia Sapega.

The case clearly did not involve any unfair prosecution against the Pilot in Command, who had no option but to comply with the instruction in order to meet their obligation for the safety of passengers in the light of a bomb threat; while Belarus clearly had violated Article 4 of the Chicago Convention. The downstream consequence of the criminal proceedings against Protasevich, however, necessitates some comment on forced confessions which may be very relevant to aviation professionals in the future.

The day after incident, Belarusian authorities issued a video in which Protasevich claimed he had been treated correctly and not been harmed and, a month later, he was paraded on State television in an interview in which he confessed to organising mass riots. The circumstances surrounding his confession gave rise to suspicions of a forced confession, whether by violence, inducement or intimidation, which demand some attention. The admissibility of such evidence may become very relevant in a case in which an aviation professional is placed in the position of such incrimination, which has a direct effect on their human rights due to its compromise of a fair trial.

In the UK, a confession is defined under Section 82 Police and Criminal Evidence Act 1984 as *any statement wholly or partly adverse to the person who made it, whether made to a person in authority or not and whether made in words or otherwise.* Section 76 enables a confession made by the Accused to be given in evidence at their Trial provided that it is relevant and admissible, but that

> if, in any proceedings where the prosecution proposes to give in evidence a confession made by an accused person, it is represented to the court that the confession was or may have been obtained –
>
> a. by oppression of the person who made it; or
> b. in consequence of anything said or done which was likely, in the circumstances existing at the time, to render unreliable any confession which might be made by him in consequence thereof, the court shall not allow the confession to be given in evidence against him except in so far as the prosecution proves to the court beyond reasonable doubt that the confession (notwithstanding that it may be true) was not obtained as aforesaid.

The consequence is that not only will the confession be excluded but also that any facts discovered as a result of the confession will be excluded from evidence at the Trial.

The use of the word *Oppression* inevitably invokes Article 3 of the European Convention on Human Rights which provides that *No one shall be subjected to torture or to inhuman or degrading treatment or punishment.* The Crown Prosecution Service qualifies this in guidance,[21] following a 1987 case,[22] that the word was to be given its ordinary dictionary meaning and was likely to involve some impropriety on the part of the interrogator. This would include a confession obtain as the result of an inducement – for example a promise of bail or a promise that a prosecution would not arise from the confession.

The question thus arises, whether a statement made during an investigation by the Air Accident Investigation Branch, would amount to a confession, that subsequently would incriminate the individual in criminal accountability. While the aviation professional is compelled to co-operate in the investigation process, the issue arises as to whether there is thus a conflict with the individual's human right against self-incrimination, which would render such a confession inadmissible.

Case Study: Hoyle v Rogers[23]

In the 2014 decision in *Hoyle v Rogers*, the Court of Appeal upheld the lower court's decision that the AAIB report, containing all the opinions and analysis, could now be admitted in proceedings, giving rise to three issues:

In accordance with normal rules of evidence, the maker of a report ordered to be admitted under UK law in Part 32 of the Civil Procedure Rules; must be available for cross-examination; otherwise, the robustness of assertions of fact or opinion contained in the report would be undermined as unsafe. The effect is that the confidentiality of the investigators contributing to the report may be more cautious about the statements they make, diluting their contribution to air safety.

As we have seen, such opinions which could lead to a Court's finding of negligence in civil proceedings may subsequently underpin a decision to arraign a Defendant on charges underpinned by the emergence of criminal negligence.

The downstream consequence is that interviewees in the investigation process may be reluctant to co-operate with the investigating authorities and, despite State regulations compelling their co-operation, would be justified by their privilege against self-incrimination, entitling them to be cautioned before an interview and to have independent legal advice throughout the interview, so that they may have counsel on their right to silence. The further dilution of the value of such investigations to air safety is obvious.

European Regulation No. 376/2014 has been vaunted as a guardian of the rights of individuals against such abuse but, even at such a delicate stage in its nascent development, its value in this context must be cast in doubt. Article 15.2 provides that information derived from occurrence reports shall be used only for the purpose for which it has been collected and that Member States *shall not make available or use the information on occurrences in order to attribute blame or liability the Article adds the provision or for any purpose other than the maintenance or improvement of aviation safety.* The case of *Hoyle v Rogers* puts such a noble provision firmly in its place. Most pertinently, Article 16 provides in paragraph 10 that the protection afforded to individuals shall not apply *in cases of wilful misconduct* or *where there has been a manifest, severe and serious disregard of an obvious risk and profound failure of professional responsibility to take such care as is evidently required in the circumstances, causing foreseeable damage to a person or property, or which seriously compromises the level of aviation safety.*

This does nothing more than underline the principle that a State's criminal law shall remain inalienable within its sovereign jurisdiction. There can be no qualification of this, so the mischief of criminal negligence and its uncertain application between different regimes remains unaltered. However unjust it may seem to the rest of the world, the rest of the world cannot do anything about it, thus perpetuating the criminalisation phenomenon, even if it is inconsistent with the body of international understanding as perceived in the Just Culture, which we will examine in more detail. This concept has received the unimpeachable authority of the European Court of Human Rights which upheld in the case of Captain Mangouras the right of the prosecuting State to deprive the individual of his human rights, notwithstanding any evidence of blame under the criminal law (which in his case necessitated a second Trial before the State got the verdict it wanted).[24]

It is inevitable that any discussion or analysis of a 'Just Culture' that ignores the gorilla of gross negligence sitting in the corner will merely perpetuate the problem of criminalisation which, ultimately, would defeat the whole purpose of the policy to protect responsible professionals who are trying to do their job.

When is a Defendant likely to be successful in arguing this privilege? The relevant principles in relation to the grant of a stay where there are related proceedings were summarised by Gloster J as follows:[25]

- It "is a power which has to be exercised with great care and only where there is a real risk of serious prejudice which may lead to injustice."
- The discretion must be exercised by reference to the competing considerations between the parties; the court must balance justice as between the two parties; a claimant has a right to have its civil claim decided; the burden lies on a defendant to show why that right should be delayed.

- A defendant must point to a real, and not merely notional, risk of injustice.
- A defendant has a right to remain silent in criminal proceedings, and would, by serving a defence in civil proceedings, be giving advance notice of his defence, carries little weight in the context of an application for a stay of civil proceedings.
- A defendant is expected to adumbrate a positive defence at an early stage in criminal proceedings, therefore, the disclosure of a defence in civil proceedings is unlikely to disadvantage a defendant in criminal proceedings.
- A positive defence is likely to exculpate, rather than incriminate, a defendant.
- It is not enough that both the civil and criminal proceedings arise from the same facts, or that the defence of the civil proceedings may involve the defendant taking procedural steps such as exchanging witness statements and providing disclosure of documents which might not be imposed upon them in the criminal proceedings.
- A defendant has a choice between remaining silent in the civil proceedings or risk giving an indication of his defence which may be used by the prosecuting authorities, the harshness of such a choice does not provide a good ground for staying civil proceedings.
- Whether safeguards can be imposed in respect of the civil proceedings which provide sufficient protection against the risk of injustice.

We should take this very seriously indeed when discussing Just Culture in the next chapter.

Inevitably the issue takes a step further if the individual is led to believe that what they say will not be used in criminal proceedings, thus rendering the statement inadmissible and in conflict with the individual's human right against self-incrimination, which would render such a confession inadmissible.

Case Study: Captain Schröder of the *Zim Mexico III*

In the United States, the Seaman's Manslaughter Act provides:

> Every captain, engineer, pilot or other person employed on any steamboat or vessel, by whose misconduct, negligence, or inattention to his duties on such vessel the life of any person is destroyed, and every owner, charterer, inspector or other public officer, through whose fraud, neglect, connivance, misconduct, or violation of law the life of any person is destroyed, shall be fined under this title or imprisoned not more than ten years or both.[26]

On 1 March 2006, Captain Wolfgang Schröder was in command of the containership *Zim Mexico III*, entering the Port of Mobile, Alabama. There were no special hazards or issue brought to mariners' attention and, the following day, having completed cargo operations she was under way for departure. The vessel encountered difficulties completing the manoeuvre in the river, which expert evidence attributed to bow thruster failure, which resulted in an allision when the bow of the ship hit a dockside crane, causing it to collapse and crushing a shoreside fitter, Shawn Jacobs, who died of his injuries.

Captain Schröder was arraigned on an indictment under the Seaman's Manslaughter Act in which the Prosecution alleged the negligence or inattention to duties of Captain Schroeder resulted in the death of Mr Jacobs.

During pre-trial case management proceedings, the Judge ruled that the standard of proof under the statute was simple negligence, in other words the civil burden was required, rather than the minimum standard required for criminal accountability of gross negligence.

The Jury even wrestled with this point in their deliberations, and sent two notes to the Judge, seeking clarification and asked if they could add some text to the verdict that they found the Master Guilty of simple negligence rather than gross negligence. The Judge directed them to write Guilty with no further words of explanation.

Captain Schröder was duly convicted and was remanded in custody for sentencing. At the sentencing hearing, some four months later, Judge Granade got as close as she dared to an apology for the injustice of the law, when she said, *While I certainly do not discount the terrible consequences that have resulted from this negligence, what he has been convicted of is really a civil offense.*

This maritime case has an uncomfortably close relevance to civil aviation, as it addresses the admissibility of what may amount to a confession, during the process of an investigation.

At the Trial, the Prosecution relied on evidence in the casualty investigation, which, as with all casualty investigations, was not gathered with a view to being given in criminal proceedings and, as a result, Captain Schröder was not cautioned at the time he was interviewed. Indeed, Counsel for the vessel's owner testified at the Trial that he had asked the Coast Guard investigator if this were a criminal investigation and was told emphatically *No*. Having been reassured on this, the Master was quite content to proceed with the interviews in the absence of a lawyer to advise him; given the Coast Guard's assurance, there was no danger that Captain Schröder might say something that would incriminate himself in a future prosecution. Then, of course, he was charged. At the Trial, the Coast Guard special agent gave evidence disputing that he had said *No*, arguing that he had said, *Not yet*. In fact, it is apparent that the Coast Guard had already prepared its criminal case before interviewing the Master again, after which they were able to refine the charge in the indictment to improve the prospect of a conviction. If nothing else, the investigation evidence merely would have addressed civil liability in negligence, and it would have been a flagrant disregard of his rights to deceive or lull him into incriminating himself in the civil context when he was under criminal investigation for his activities.

Despite this, the Court refused to exclude the evidence of the statements apparently conducted after misrepresenting the situation to Captain Schröder, and the statements went before the Jury, who would have been entitled to draw a conclusion that there had been some inconsistency in Captain Schröder's position.[27]

This clearly gives rise to the admissibility of evidence obtained unfairly; Section 78 of the 1984 Act states:

> In any proceedings the court may refuse to allow evidence on which the prosecution proposes to rely to be given if it appears to the court that, having regard

to all the circumstances, including the circumstances in which the evidence was obtained, the admission of the evidence would have such an adverse effect on the fairness of the proceedings that the court ought not to admit it.

This Judge, after all, has the essential function of ensuring a fair trial that meets the requirements of the European Convention on Human Rights.

BACK TO BASICS

It is glaringly obvious that we have strayed a long way from the basis of a prosecution, to apply the evidence to the elements defined in the law in question, and to do so fairly, in respect of the Defendant's human rights. In the context of civil aviation, the evidence upon which they rely for a conviction will necessarily be found in the investigation process, as a result of which they will seize upon any evidence that they can find, notwithstanding that such evidence may have been obtained during the air accident investigation in which the Defendant may have co-operated in the confidence that they would not be incriminating themselves. Naturally, such a privilege would not serve the purposes of the prosecution (or the Courts) at all, and the result creates confusion in separating the accident investigation, conducted on a scientific methodology with a view to preventing future accidents, from the criminal investigation to pursue a criminal conviction, which must respect the procedural rules and the human rights of the Defendant.

Paradoxically, the Chicago Convention creates the mischief itself. Further, 3.1 of Annex 13 emphasises that *the sole objective of the investigation of an accident or incident shall be the prevention of accidents and incidents. It is not the purpose of this activity to apportion blame or liability.* In order to achieve this, 3.2 states that *A State shall establish an accident investigation authority that is independent from State aviation authorities and other entities that could interfere with the conduct or objectivity of an investigation.* However, Article 12 states: *Each contracting State undertakes to insure the prosecution of all persons violating the regulations applicable.* The result is a blurring of the separation between investigation for the prevention of future accidents, and prosecution of an identified offender.

The CAA is tasked by the Department for Transport to investigate and prosecute breaches of aviation safety rules and some aviation related consumer protection and health and safety requirements. Certainly the UK has observed the concept, with the establishment of the Air Accidents Investigation Branch (AAIB), which investigates civil aircraft accidents and serious incidents within the UK, its overseas territories and crown dependencies. As we have seen with the case of *Hoyle v Rogers*, however, the primary objective does not prevent a Court admitting the report in evidence. EU regulation 996/2010 requires that certain records *shall not be made available or used for purposes other than safety investigation* which includes all statements taken from persons by the safety investigation authority. Naturally this would afford some comfort to the investigators, who thus would be spared the ordeal of being cross-examined in evidence, which might shake their confidence somewhat. The AAIB's investigations are entirely independent from any industry or judicial investigations, and must not at any time disclose to anyone else, including the police, the Civil Aviation Authority, the interviewee's employer or any other enforcing authority, personal details in connection with any information contained in their statement, without a Court Order, as we have seen. But this becomes blurred when taking into account a Memorandum of Understanding with the Crown Prosecution Service. A brief summary of this MoU will be helpful.

It is not legally binding, and seeks to explain that its aim is to create practical working arrangements between the accident investigation branches (AIBs) and the CPS. Having acknowledged the AIBs' duty to conduct safety investigations into accidents in their respective transport sectors without apportioning blame or liability, it then states:

> Where there is evidence that an offence has been committed and it meets the Full Code Test in the Code for Crown Prosecutors, the CPS has a duty to make sure that the right person is prosecuted for the right offence, and to bring offenders to justice wherever possible.

This appears to infer that evidence revealed by the investigation would be used to prosecute an individual. Accordingly, the MoU provides guidance to facilitate cooperation (while respecting the AAIBs' independence) with the CPS. Some basic principles are defined:

- Safety investigations and criminal investigations will proceed in parallel. The aim of this MOU is to ensure that both investigations can progress effectively and independently, while promoting cooperation where permitted.
- The interests of an AIB investigation may sometimes take precedence over a criminal investigation. The ability of witnesses to talk openly and without delay to an accident investigator is fundamental to the operation of the AIBs, as is the protection of sensitive material from disclosure.
- In criminal proceedings, the CPS must consider whether evidence can be used in court and is reliable and credible, and ensure there is no other material that might affect the sufficiency of evidence. The AIBs will consider all requests from the CPS for specific items of evidence and factual information and will disclose such material where permitted by their respective legal regimes.
- AIBs are prohibited from disclosing witness details, statements or declarations unless ordered to do so by the High Court.
- Generally, witness statements provided to the Air Accidents Investigation Branch (AAIB) cannot be disclosed in the absence of a court order. BUT
- Where the AIBs are generally prohibited from disclosing information, its disclosure can be ordered by the High Court. The court is required to carry out a public interest balancing test between the benefits of disclosure in the interests of justice against the potential adverse impact it might have on the AIBs in conducting their current or future investigations.
- An AIB may commission reports from experts on technical or other specialist matters relevant to its investigation. AAIB experts' reports cannot be disclosed unless the High Court orders disclosure.

Annex B of the MoU identifies some areas of the report for which disclosure cannot be made without a Court Order:

- Statements
- Identities of witnesses to the event
- Medical records
- Draft reports
- Records revealing the identity of persons who have given evidence in the context of the safety investigation including third party experts
- Cockpit video and image recordings and transcripts, air traffic control recordings, flight data recorders
- All communications between persons having been involved in the operation of the aircraft
- Information collected by the AAIB which is of a particularly sensitive or personal nature.

Some factual information is subject to discretionary disclosure by the AAIB, relating to factual material which does not fall under Article 14[28] of EU Regulation 996/2010 and proceedings of the investigation may be disclosed provided that their disclosure does not compromise the objectives of the investigation. At first sight, the watertight protection does not seem quite so watertight when compared with the basics of the individual's human rights, such as the respect for private life (medical records, information of a sensitive or personal nature) but Article 8 of the Convention provides the excuse:

> There shall be no interference by a public authority with the exercise of this right except such as is in accordance with the law and is necessary in a democratic society in the interests of national security, public safety or the economic well-being of the country, for the prevention of disorder or crime, for the protection of health or morals, or for the protection of the rights and freedoms of others.

No mention is made, however, of the right against self-incrimination.

In her monograph on the criminalisation of air accidents,[29] Mildred Trögeler conducted a comparative study of the international and domestic frameworks of relevant law and concluded that there were significant areas in which the two suffered conflicting interests, with the result that there was no single regime to be applied across international and domestic regimes. The theory is a sound one, because transnational concepts of human rights can only be put into practice through domestic laws but, as we have seen, the criminal process is jealously guarded by the sovereign jurisdiction of domestic courts, which presents a conflict of priorities with international provisions for the investigation of air accidents and, as we shall see, Just Culture. Her point is very relevant to the investigation process, rather than the State's legal process, with the need to adopt more stringent rules across States which balance the interests of both the technical investigations, at the heart of international law, and the criminal investigations which drive prosecutions in domestic courts.

At the heart of the basics, it is all well and good to proclaim that the administration of criminal law is the sovereign domain of State society, which cannot defer to any contrary demands in an international convention, but the political motivations which underpin State laws clearly risk the aviation professional's human rights, which are defined by international convention; the result is that an unstoppable force meets an immovable object. In reality, of course, the criminalisation process will be driven by the domestic agenda, and the normative ethics which delivered convictions against Captain Mangouras and Captain Chafik. In order to meet the political agenda, the dilemma can be a compelling one.

Case Study: The *Concorde* crash

On 25 July 2000, a Concorde on Air France Flight 4590, crashed two minutes after take-off from Paris Charles de Gaulle Airport, killing all 109 people on board and four on the ground. It was the only fatal Concorde accident during its 27-year operational history but would have consequences that ultimately brought the entire Concorde dynasty to an end.

During its take-off, the aircraft ran over debris on the runway, causing a tyre to explode and disintegrate. Tyre fragments, launched upwards at great speed by the rapidly spinning wheel, violently struck the underside of the wing, damaging parts of the landing gear, preventing its retraction, and causing the integral fuel tank to rupture. The ruptured fuel tank haemorrhaged fuel which rapidly ignited, causing a loss of thrust in both port-side engines. The aircraft actually began to lift off, but the catastrophic damage rendered the aircraft impossible to control, crashing into a building.

Paradoxically, the villain of the piece was not Concorde, but a Continental Airlines McDonnell Douglas DC-10, which earlier had taken off from the same runway when it lost a titanium alloy strip that was part of the engine cowl. The Concorde ran over this piece of debris during its take-off run while the aircraft was doing 185 miles per hour, cutting the right-front tyre of its left main wheel bogie and sending a large chunk of tyre debris into the underside of the left wing at an estimated speed of 140 metres per second. It did not actually puncture the fuel tank, but sent out a pressure shockwave that ruptured the tank at its weakest point, and leaking fuel was most likely ignited either by an electric arc in the landing gear bay or through contact with hot parts of the engine. By the time that the flight crew were alerted to the fire, it was too late to abort take-off, which would have led to a high-speed runway excursion and collapse of the landing gear, which also would have caused the aircraft to crash. The flight crew tried to divert to nearby Paris–Le Bourget Airport, but it was too late.

The investigation process was commenced, according to French procedure, by *the Bureau d'Enquêtes et d'Analyses pour la sécurité de l'aviation civile*, or BEA. A criminal investigation was opened at the same time by the French judicial authority, and mutual co-operation was required under ministerial instructions that had been defined in 1953, so was well rehearsed. The concept was designed for two levels, with the BEA establishing the technical facts, which the judicial authority would then adopt in order to complete their work.

Naturally, such an operation relied on mutual trust; but that trust had been broken in 1988, when the BEA investigated a crash near Mulhouse and duly took the flight recorders, but when they were later handed over to the criminal investigators, a few seconds recorded prior to the crash were missing, giving rise to the serious allegation that the BEA had tampered with the evidence in order to conceal a defect in the aircraft. With the loss of trust, the ministerial instructions were abandoned and the judicial investigation would now take precedence over the BEA. Trögeler[30] astutely unravelled the key issues in French law that now arose; given the loss of trust, French law authorises the judicial investigation to take the flight recorders and then make them available to the BEA who can make copies under the supervision of a police officer.

Annexe 13 of the Chicago Convention contains general principles to be applied in accident investigations:

> 3.3 The State of Occurrence shall take all reasonable measures to protect the evidence and to maintain safe custody of the aircraft and its contents for such a period as may be necessary for the purposes of an investigation. Protection of evidence shall include the preservation, by photographic or other means, of any evidence which might be removed, effaced, lost or destroyed. Safe custody shall include protection against further damage, access by unauthorized persons, pilfering and deterioration.

In consequence, therefore, the French process of giving priority safe custody to the judicial investigation is not offensive, in order that the balance be struck between aviation safety and domestic law. However, there is merit in Trögeler's argument that the process is intended to prioritise the objective in French law to conduct the investigation with a view to meeting the objective of criminal accountability for the purpose of satisfying the rights of alleged victims and their dependents, who have the right to pursue criminal proceedings. Whose human rights should therefore have priority? The judiciary clearly intend that the objective of criminal accountability should be upheld; which holds well until 5.4 of Annex 13 is considered:

> The accident investigation authority shall have independence in the conduct of the investigation and have unrestricted authority over its conduct, consistent with the provisions of this Annex. The investigation shall normally include
>
> a. the gathering, recording and analysis of all relevant information on that accident or incident;
> b. the protection of certain accident and incident investigation records in accordance with 5.12;
> c. if appropriate, the issuance of safety recommendations;
> d. if possible, the determination of the causes and/or contributing factors; and
> e. the completion of the Final Report.

5.6 takes us a little further:

> The investigator-in-charge shall have unhampered access to the wreckage and all relevant material, including flight recorders and ATS records, and shall have unrestricted control over it to ensure that a detailed examination can be made without delay by authorized personnel participating in the investigation.

The conflict is obvious. Both investigations state that their sole aim is the prevention of future accidents and not to apportion blame or liability, but clearly the judicial investigation is intended to identify and apportion blame according to the evidence, and has the power to hamper access to the BEA of evidence which may lead to their disadvantage. Some third party provision would be helpful in resolving the conflict; this is readily given by Article 6 of the 1998 European Directive,[31] establishing the fundamental principles governing the investigation of civil aviation accidents and incidents provides:

> 1. Each Member State shall ensure that technical investigations are conducted or supervised by a permanent civil aviation body or entity. The body or entity concerned shall be functionally independent in particular of the national aviation authorities responsible for airworthiness, certification, flight operation, maintenance, licensing, air traffic control or airport operation and, in general, of any other party whose interests could conflict with the task entrusted to the investigating body or entity.
> 2. Notwithstanding paragraph 1, the activities entrusted to this body or entity may be extended to the gathering and analysis of air safety related data, in particular for prevention purposes, in so far as these activities do not affect its independence and entail no responsibility in regulatory, administrative or standards matters.
> 3. The body or entity referred to in paragraph 1 shall be given the means required to carry out its responsibilities independently of the authorities referred to in paragraph 1 and should be able to obtain sufficient resources to do so. Its investigators shall be afforded status giving them the necessary guarantees of independence. It shall comprise at least one investigator able to perform the function of investigator-in-charge in the event of an aircraft accident or serious incident.

It is noteworthy that we have not yet encountered any reference to the CAA's Investigations and Enforcement Team (IET) in the UK. Notwithstanding anything above, the CAA has the power to prosecute companies and individuals for breaches of the legislation it is tasked with enforcing. Investigations into alleged breaches of the law are carried out by the IET, in accordance with its own Code of Practice. Its policy is stated as being complementary to its commitment to a Just culture approach to safety regulation but empowers it to take action *where necessary* to address non-compliance with the regulations.

THE PLACE OF HUMAN RIGHTS IN THE IET'S LEGAL FRAMEWORK

The Legal Framework is comprehensively set out in its Code of Practice.[32]

Aviation safety legislation applicable in the UK includes EC Regulations and Directives, the Air Navigation Order 2016 (ANO), the Aviation Security Act 1982, Regulations made under the ANO and its predecessors, such as the Rules of the Air Regulations 2015, and the Air Navigation (Dangerous Goods) Regulations 2002. Article 265 of the ANO provides that it is a criminal offence to contravene EC Regulations, the ANO or Regulations made under the Order. There are four categories of offence:

1. The most serious offence is that set out in Part 4 of Schedule 13 ANO for which the maximum penalties are set out in Article 265(8) ANO, namely on summary conviction in England and Wales an unlimited fine, or in Northern Ireland or Scotland a fine not exceeding the statutory maximum (currently £5,000 in Northern Ireland or £10,000 in Scotland). On conviction on indictment, the maximum penalty is an unlimited fine and/or imprisonment for five years.
2. The next most serious category of offence is that listed in Part 3 of Schedule 13 ANO. The penalties are set out in Article 265(7) ANO, namely on summary conviction in England and Wales an unlimited fine, or in Northern Ireland or Scotland a fine not exceeding the statutory maximum (currently £5,000 in Northern Ireland or £10,000 in Scotland). On conviction on indictment, the maximum penalty is an unlimited fine and/or imprisonment for two years.
3. Next are those offences listed in Part 2 of Schedule 13 ANO. These offences can only be tried in the Magistrates Court in England, Wales or Northern Ireland, or in the Sheriff Court in Scotland. The penalty is at Article 265(6) ANO and the maximum penalty is a fine not exceeding level 4 on the standard scale (currently £2,500).
4. The final category of offence is anything which is listed in Part 1 of Schedule 13 ANO. Again, these can only be tried in the Magistrates Court in England, Wales or Northern Ireland, or in the Sheriff Court in Scotland. For these offences the maximum penalty is at Article 265(5) ANO and is a fine not exceeding level 3 on the standard scale (currently £1,000).

Where a case may be heard in either the Magistrates or the Crown Court, the Magistrates decide whether they feel able to deal with it. The decision is based on the seriousness of the charge, the legal complexity of the case and whether in the event of conviction it is likely that a penalty greater than the maximum available in the Magistrates Court may need to be imposed. A defendant may demand that the case be referred to the Crown Court.

Protection of the individual's human rights is given in some detail, notably:

- Service standards. IET staff will
- Be courteous and helpful.
- Respond to calls and correspondence promptly and in any case within 20 working days.
- Prioritise their workload proportionate to the risks as well as in accordance with our statutory and regulatory responsibilities.
- Ensure that our investigative and decision-making actions are appropriately documented and recorded.
- Treat those involved in or subject to such actions (whether individuals or organisations) in a consistent manner.
- Explain clearly the consequences of non-compliance with the IOs' requests and the investigation's requirements.
- Ensure that individuals or organisations under investigation are made aware of the allegations.

- Ensure that individuals or organisations under investigation are given an opportunity to respond to the allegations against them.
- Ensure that witnesses and victims are kept informed of the progress of an investigation.
- The interview process

Interviews are always voluntary, and the individual subject of the interview will always be offered the opportunity to take legal advice at their own cost. Interviews, with some restrictions, can take place anywhere convenient to the person being interviewed. … Where the person under investigation may be prosecuted the interview will be conducted under caution and be recorded. The person under investigation will be given adequate opportunity to clarify the nature of the offence and to respond to specific allegations put to him or her. They will also be asked to confirm that he or she has received pre-interview disclosure.

THE IET'S PROCEDURE

Reports and allegations concerning breaches of the law and regulations are sent in to the IET, and other departments in the CAA, by members of the public, the aviation community, internal CAA departments and law enforcement agencies. Such reports are referred to the appropriate department in the CAA (such as flight operations, air traffic management, airworthiness etc.) for a decision as to whether or not they should be referred to the IET for investigation. On referral of these reports and allegations to IET, they are reviewed by the IET manager or team leader. If there is credible and verifiable evidence indicating a breach of regulations which is a serious risk or is detrimental to consumer protection then an IO[33] will be appointed to undertake initial enquiries. These initial enquiries may involve investigation and inquiry with the original complainant to obtain further information and/or documentation. Some key issues follow in the procedure:

- If the information continues to demonstrate a potential breach of regulation, and the evidence appears reliable, truthful and likely to be substantiated by other evidence, the IO will establish contact with the person under investigation, if he/she has not already done so.
- In the course of their investigation IOs will take written witness accounts, secure copies of documents and formally interview alleged offenders.
- The provision of information and documentation to the person under investigation will be recorded by the IO in a case file with details as to dates, times, any persons present etc.
- In every case the investigating officer will indicate the alleged offence.
- Once an investigation is completed, the case file will be sent to a CAA Legal Adviser for an assessment in accordance with the Code for Crown Prosecutors.
- The IO, or Support Officer on behalf of the IO, will inform witnesses and the person under investigation when they submit their report for legal analysis. Thereafter, the Legal Adviser is responsible for ensuring that witnesses and the person under investigation are kept informed of progress by the IO.

- Having assessed the evidence, the Legal Adviser will recommend a case outcome to the relevant CAA department.
- Outcomes include no further action, a warning, a Simple Caution or a Conditional Caution or prosecution. Simple Cautions and Conditional Cautions are given only where a person has admitted the offence(s). They do not amount to a criminal conviction but will remain on a person's CAA record and can be taken into account by the Courts if a person is subsequently convicted for a further offence.
- At any time during an investigation, the IET may refer cases to the relevant department within the CAA for licensing action in accordance with the Principles set out above. Such regulatory action is taken independently from any prosecution.

THE PROBLEM OF SEPARATION OF POWERS

It is therefore apparent that the ancient principle of a separation of powers has become blurred. The CAA makes the Regulations, then investigates relevant breaches and determines the outcome. UK's Supreme Court quotes the French social commentator and political thinker Montesquieu, who stated in 1748

> When the legislative and executive powers are united in the same person, or in the same body of magistrates, there can be no liberty…there is no liberty if the powers of judging is not separated from the legislative and executive… there would be an end to everything, if the same man or the same body… were to exercise those three powers.

Notwithstanding the provisions nobly enshrined in the Constitutional Reform Act 2005, the human rights of the individual appear to have been eroded by the powers extended to the CAA, even though the prosecution function has already been established to run parallel with the CPS, as we have seen.

Human rights are enshrined in international law, but the enforcement process demands that all the domestic laws of a State must be complied with; so for example the air transport licence must be in place and valid for the jurisdiction in which the flight takes place; it matters not that a licence may have been issued in another State.

Section 64 (1) Civil Aviation Act 1982 particularly prohibits unlicensed carriage of passengers or cargo unless:

a. the operator of the aircraft holds a licence granted to him by the CAA in pursuance of section 65 below (in this Act referred to as an "air transport licence") authorising him to operate aircraft on such flights as the flight in question; and
b. the terms of the licence are complied with so far as they relate to that flight and fall to be complied with before or during the flight.

Subsection 5 states that, where an aircraft is used on a flight in contravention of subsection (1) then a person guilty of an offence shall be liable on

a. summary conviction, to a fine not exceeding the statutory maximum; and
b. conviction on indictment, to a fine or to imprisonment for a term not exceeding two years or to both.

Case Study: The Abbotsford Crash and a Can of Worms

In this respect, the case of a crash in Canada in August 2018 is noteworthy. The casualty itself was not caused by the negligence of the pilot or the operator. In a Transportation Safety Board Investigation Report issued on 28 March 2019, investigators cited a strong crosswind as a possible factor that led to the crash of an American-operated vintage de Havilland DH-89A Dragon Rapide shortly after take-off during the Abbotsford International Airshow at British Columbia's Abbotsford airport. The aircraft had been part of a static display but, following the conclusion of the airshow, the aircraft was being used to provide tourist rides for fare-paying passengers; this was by no means unauthorised by the operator, for the pilot was the operator's founder.

At approximately 17.31, the DH-89 began its take-off with the pilot and four passengers on board. Air traffic controllers had advised the pilot that winds were 180 degrees gusting 15 knots. Runway 19 was available, but aircraft displays occupied much of the runway's length, so the pilot decided to use Runway 25 instead. The crosswind component was calculated to be 18 knots at the moment of take-off. The DH-89 encountered gusting winds approximately 30 feet above the runway and lost airspeed, causing it to nose over and crash just beyond the edge of Runway 25. The pilot and one passenger were seriously injured, and three other passengers sustained minor injuries.[34]

The aircraft was owned and operated by a US company. The pilot was highly competent, having held a valid American airline transport pilot licence and a valid medical certificate, which also allowed the pilot to fly in Canada. He had accumulated about 10,000 total flying hours over a 35-year period on many types of aircraft, including 2,500 hours on tail-wheel-equipped aircraft. He was endorsed to operate a variety of vintage aircraft. He had obtained flight instruction from a world expert on the Dragon Rapide, and had accumulated about 20 hours total time on type, including about five hours on the occurrence aircraft in the month prior to the accident. He had also conducted another flight with passengers on the morning of the accident, so was well-familiarised with the aircraft; even so, the vintage DH-89 was not an easy aircraft on novice pilots; even the Canadian casualty report stated:

> Historical information indicates that the Dragon Rapide is more difficult to handle during crosswind take-offs than most tail-wheel-equipped aircraft, particularly from paved (rather than grass) runways. According to pilot notes, the maximum allowable crosswind component for take-off is 17 knots. This is not a limitation of the aircraft; rather, it is a cautionary speed, meaning that operating in winds above that value would require above-average flying skills.

The higher level of caution needed necessarily demanded a higher duty of care, but there was no evidence to suggest that the pilot had failed to reach the threshold required of any pilot in the circumstances of this case. Indeed, the evidence would have led to the conclusion that he had done everything that might reasonably have been expected of him; the issue of the crosswind mentioned in the casualty report was also raised in the crash of a Dragon Rapide which crashed at Duxford in June 2022 when there was a significant crosswind from the left, although it was within the aircraft limit (stated as 20 knots).[35] In neither case had there been any fatality. In both cases, there would have been no just cause to initiate criminal proceedings. In the case of the Canadian casualty, however, an unexpected issue was raised, in the certification of the aircraft, that would certainly have led to much concern.

In the United States, the Federal Aviation Administration (FAA) has the power to issue a Special Airworthiness Certificate – Experimental which allows the operation of vintage and other aircraft that do not qualify for a normal Certificate of Airworthiness. These aircraft are subject to multiple restrictions and limitations, including a prohibition from carrying passengers for hire or reward. However, the FAA has a policy in place that allows the owners of certain vintage aircraft to apply for an exemption that would permit the carriage of passengers for reward. The aircraft in question did not have one, however. The Canadian Aviation Regulations prohibit aircraft that have been issued a Special Certificate of Airworthiness from carrying persons for hire or reward. Transport Canada can grant an exemption to commercial air service operators to perform aerial work while carrying passengers on board for hire or reward in Canadian-registered historical aircraft; however, this exemption does not apply to American-registered aircraft, and the operators were not considered a commercial air service operator in any event. That being said, in certain circumstances, Transport Canada may grant an exemption to an aircraft that holds a Special Certificate of Airworthiness; however, the aircraft requires an exemption from the issuing State first and, of course, the operators had not obtained one.

Transport Canada had issued a special authorisation (Foreign Flight Authority) to the operators to fly the DH-89 into Canada, but that did not allow for the carriage of passengers for hire or reward. It can well be understood, however, that the application of Section 64 of the UK's 1982 could have led to a criminal prosecution.

Whether or not a criminal prosecution was even contemplated following the accident, the downstream consequence of the licensing failure would likely have led to insurance issues, which might have been even more serious, for the civil claims which might have been contemplated from the injured passengers would naturally have been transferred to the insurance underwriters. Anybody who would have benefitted by the safety or due arrival of the insured property, or would be prejudiced by its loss or damage, may have an insurable interest. Therefore, apart from the insurable interest that would naturally have arisen for the owners of the aircraft, the small print in the policy might have limited or, indeed, excluded liability in the event that the critical certification had not been carried out in the State in which the accident occurred.

At the very least, it provides a stark illustration of how the pilot in command bears the responsibility for the safety of life, safe flight and risk management of the operation in an emergency, which exposes them to all manner of civil and criminal accountability, whenever an element of negligence arises, and the Just Culture must struggle to make itself heard.

NOTES

1 July 2023.
2 The right to a fair trial.
3 Re Medicaments and related Classes of Goods (No. 2) [2001] 1 WLR 700.
4 The defendant was convicted at the retrial.
5 *Mangouras v Spain* (12050/04). The author's reading of the text of the Judgment, which is in French.
6 Aaronson, E and G Shaffer, 2021, Defining Crimes in a Global Age: Criminalization as a Transnational Legal Process. *Law & Social Inquiry* 46 (2) (May 2021) 455–486.
7 *R v Adomako* [1995] 1 AC 171.
8 *Northern Securities Co v United States* 193 US 197 (1904).
9 See, e.g., Powell J and J Paterson, 2004, *Cicero The Advocate*, Oxford University Press, Oxford, pp. 61–78.
10 *Donoghue (or McAlister) v Stevenson* [1932] All ER Rep 1; [1932] AC 562.
11 GM1 NCC.GEN.106 Pilot-in-command responsibilities and authority.
12 Civil Aviation Authority, 2020, CAP 1422.
13 *Eweida v United Kingdom* [2013] ECHR 37.
14 *Mangouras v Spain* (12050/04).
15 *Mangouras v Spain* (12050/04); Para 3 Summary of Judgment.
16 www.ifalpa.org/downloads/Tuninter.pdf, accessed 25 September 2015.
17 SERA 1101 (General).
18 Manual on the Regulation of International Air Transport, 2004, International Civil Aviation Organization (Doc 9626, Part 4).
19 ICAO, 2022, Event Involving Ryanair Flight Fr4978 in Belarus Airspace on 23 May 2021.
20 United Nations Security Council, 9175th meeting, 31 October 2022.
21 See www.cps.gov.uk/legal-guidance/confessions-unfairly-obtained-evidence-and-breaches-pace.
22 *R v Fulling* [1987] 2 All ER 65.
23 *Scott Hoyle v Julia Mary Rogers and Nicola Lucinda Rogers and Secretary of State for Transport (1st Intervener) and International Air Transport Association (2nd Intervener)* [2014] 3 WLR 148; [2014] CP Rep 30; [2014] EWCA Civ 257; [2014] 3 All ER 550.
24 *Mangouras v Spain* (12050/04).
25 *Akcine Bendrove Bankas Snoras v Antonov and Another* [2013] EWHC 131 (Comm).
26 18 USC Section 11153.
27 *United States of America v Wolfgang Schröder* [2007] United States District Court, Alabama Southern District (unreported).
28 Protection of sensitive safety information.
29 Trögeler M, 2010, *Criminalisation of Aviation Accidents: Creation of a Just Culture: Balancing Aviation Safety and the Proper Administration of Justice*, Leiden University Law School, Leiden.
30 Trögeler M, 2010, *Criminalisation of Aviation Accidents: Creation of a Just Culture: Balancing Aviation Safety and the Proper Administration of Justice*, Leiden University Law School, Leiden.
31 Council Directive 94/56/EC of 21 November 1994.
32 Code of Practice for the Investigations and Enforcement Team CAP 1422.
33 CAA Investigation Officer.
34 Air Transportation Safety Investigation Report A18P0115.
35 The AAIB report concluded that the aircraft suffered a hard landing and runway excursion after the right main landing gear tyre had deflated either in flight or on touchdown. AAIB-28384 (See AAIB Bulletin: 10/2022).

4 What Is the Theory of a Just Culture?

OVERVIEW

The Civil Aviation Authority (CAA) asserts that the aim of a Just Culture is *to promote continuous learning from previous mistakes and to encourage pilots to openly and freely share essential safety related information.*

Civil aviation safety is sustained by a culture which fosters and facilitates the reporting of occurrences and advances the principles of a Just Culture. It states its approach as one in which people are treated fairly and are not punished for actions, omissions or decisions taken by them that are commensurate with their experience and training. In other words, a Just Culture is intended to protect the individual's human rights and thus prevent a case of criminalisation arising in a case which previously had not been considered to be a crime. However, the CAA then qualifies its position that, in order to maintain or improve aviation safety in cases of, for example, gross negligence, wilful violations and destructive acts, further action may be necessary.

The CAA is keen that members of the civil aviation community can feel they can openly report any incidents to the CAA without fear of punishment and the information they provide will be used for safety mitigation purposes and guidance only. They then state that *Where a pilot is found negligent that regulatory action may be deemed appropriate.* It is unclear how they define 'negligent,' which is a civil tort and open to challenges of definition and application; negligence itself is not a crime, but, globally, criminal negligence has been applied in civil aviation cases that undermine this theory of Just Culture, and the provisions of guidance in current law do not assist, as seen in the Civil Aviation (Investigation of Air Accidents and Incidents) Regulations 2018 which states

> 8. The sole objective of a safety investigation which is undertaken pursuant to… these Regulations is the prevention of accidents and incidents, without the apportionment of blame or liability.

As a result, we are left bewildered as to how the findings of an investigation can be used against a witness who complies with the process. In this chapter, we endeavour to clarify the theory before applying it in practice, in the context of the process of criminalisation which has undermined the confidence of the individual in the fairness of the system.

 DOI: 10.1201/9781003489696-4

The fundamental feature of a Just Culture is to strike the right balance between the need for improving aviation safety and the recognition of the judicial system's legitimacy to investigate and prosecute a crime.

Ultimately, this chapter must rationalise three questions:

- How does a Just Culture policy serve its purpose to promote continuous learning and encourage pilots to share essential information?
- How does the policy apply its purpose to the individual's rights to fairness?
- How is the policy compromised by the criminalisation process in domestic legislation?

ORIGINS

At the heart of the matter is the blame culture which evolved in the twentieth century. It started mildly enough, in a 1937 road traffic prosecution when the driver of a van drove above the speed limit and overtook another car, but in so-doing, he struck a pedestrian and killed him. He was convicted of manslaughter but appealed; indeed, the case went to the House of Lords,[1] where Lord Atkins made an important statement on the degree of negligence required for gross negligence manslaughter:

> Simple lack of care such as will constitute civil liability is not enough: for purposes of the criminal law there are degrees of negligence: and a very high degree of negligence is required to be proved before the felony is established.[2]

In Chapter 2, we examined how the criminal justice system criminalises professional standards and behaviour in the human element in the aftermath of an accident, on the basis that the standards hitherto had fallen below the threshold of a crime but that those standards are now inadequate in the normative ethics of Society and the offender should be punished. In other words, Society interprets the conduct to demonstrate evidence of the *mens rea* and *actus reus* that leads to a conviction of the offender, re-defining the application of Lord Atkin's' test in *Andrews v DPP*, in order to rationalise the decisions in *Adomako* and *Evans*.[3]

The resultant threat of prosecution for risk management decisions, that had not previously been a crime, necessarily demands an assessment of human error, with the objective of identifying whether the decision or its contextual conduct rendered that human error a breach of the duty of care in the civil tort of negligence, which was so grossly negligent that it should be seen as criminal. Such an assessment is for the Jury to decide, and juries are notoriously capricious. Of course, the case might involve complications of discretionary judgment; while the pilot in command has the discretion to depart from the Rules of the Air in circumstances that render such departure absolutely necessary in the interests of safety, they will still be accountable. It means that they will be answerable for the judgment call which they make, in response to a multitude of components which interact in the emerging crisis. As a result, after the critical event, the individual may reflect on how their conduct mapped onto their experience and training, and what other professionals might decide. It is a natural human response to human error, and humans are, after all, fallible creatures; but if the individual fears that the downstream consequence will result in blame and, ultimately, criminal punishment, an equally human response will prioritise their own defence or, at least, mitigation, over the process of learning and improving lessons on air safety, which is the underlying function of a Just Culture.

Any margin of uncertainty could compel them to protect their position, and discourage them from co-operating in the process of making reliable findings and recommendations in the light of the accident, the cornerstone of the Chicago Convention. Annex 13 does its best to iterate the point, emphasising the importance of safety information, gathered from safety data collection and processing systems and collected solely for the purpose of improving aviation safety, and its protection from inappropriate use, *in order to ensure its continued availability, since the use of safety information for other than safety-related purposes may inhibit the future availability of such information, with an adverse effect on safety.* Inappropriate use, in violation of its purpose, naturally would weigh heavily on the reflection of the individual, thus compromising the whole purpose of investigation to prevent future accidents. The solution had to be in disarming the blame culture to a level that both the individual and the State can live with, by treading a fine line which protects the individual's rights in a criminal trial without rejecting the State's rights to enforce the laws that have evolved by the normative ethics of Society.

TO PROSECUTE OR NOT TO PROSECUTE? THAT IS THE QUESTION

The fundamental question is: *Who draws the line?* Professor Sidney Dekker[4] dissected the question into three constituent parts, which we must struggle to answer. In most regimes, it is a prosecuting lawyer who draws the line beyond which an individual will be accountable for criminal negligence in a given case; by the same token, it is the prosecutor who must persuade the Court beyond reasonable doubt that the line has been crossed; but it is difficult to establish the rules which harness the technical knowledge in applying the evidence to the tenets of criminal law, evidence and practice, without a neutral viewpoint.

Professor Dekker's point makes uncomfortable reading. The answer does not lie in the values of a Just Culture, but in the normative ethics of Society, as defined in the incarnation of the State, which appoints a prosecutor who must decide whether a person should be charged with a criminal offence and, if so, what that offence should be. They may well take expert opinion on the evidence, but it is their decision which determines the official legitimacy as to whether the line has been crossed, and then rationalised within the guidance of the Code for Crown Prosecutors.[5] Where they are satisfied on the evidence that the line has been crossed and an offence committed, the Full Code Test must be applied in an objective way, which is the nearest thing to a neutral viewpoint, by applying the laws of evidence to the offence defined in the statute, as interpreted by established cases. An awful lot of confidence is placed in a prosecutor's discretion.

The Full Code Test defines the guided steps towards the decision to prosecute, starting with the evidential stage, followed by the public interest stage. Prosecutors should only take such a decision when they are satisfied that the broad extent of the criminality has been determined and that they are able to make a fully informed assessment of the public interest. If prosecutors do not have sufficient information to take such a decision, the investigation should continue and a decision taken later.

The Evidential Stage

Prosecutors must be satisfied that there is sufficient evidence to provide a realistic prospect of conviction against each suspect on each charge. They must consider what the defence case may be, and how it is likely to affect the prospects of conviction. A case which does not pass the evidential stage must not proceed, no matter how serious or sensitive it may be.

The finding that there is a realistic prospect of conviction is based on the prosecutor's objective assessment of the admissible evidence, including the impact of any defence and any other information that the suspect has put forward or on which they might rely. It means that an objective, impartial and reasonable Jury or bench of magistrates or judge hearing a case alone, properly directed and acting in accordance with the law, is more likely than not to convict the Defendant of the charge alleged. This is a different test from the one that the criminal courts themselves must apply.[6] A court may only convict if it is sure that the Defendant is guilty.

The Public Interest Stage

In every case where there is sufficient evidence to justify a prosecution or to offer an out-of-court disposal, prosecutors must go on to consider whether a prosecution is required in the public interest. It has never been the rule that a prosecution will automatically take place once the evidential stage is met. A prosecution will usually take place unless the prosecutor is satisfied that there are public interest factors tending against prosecution which outweigh those tending in favour. In some cases, the prosecutor may be satisfied that the public interest can be properly served by offering the offender the opportunity to have the matter dealt with by an out-of-court disposal rather than bringing a prosecution.

When deciding the public interest, prosecutors should consider each of the following questions so as to identify and determine the relevant public interest factors tending for and against prosecution:

- How serious is the offence committed?
- What is the level of culpability of the suspect?
- What are the circumstances of and the harm caused to the victim?
- What was the suspect's age and maturity at the time of the offence?
- What is the impact on the community?
- Is prosecution a proportionate response?
- Do sources of information require protecting?[7]

The problem is immediately discernible: the decision as to whether the line has been crossed, concerns the interests of justice and likelihood of a conviction. As the Code states, *it is not the function of the CPS to decide whether a person is guilty of a criminal offence, but to make assessments about whether it is appropriate to present charges for the criminal court to consider.* Moreover, Annex 13 of the Chicago Convention goes out of its way to distance itself from interference in the criminal process, stating in the general principles:

> The sole purpose of protecting safety information from inappropriate use is to ensure its continued availability so that proper and timely preventive actions can be taken and aviation safety improved. It is not the purpose of protecting safety information to interfere with the proper administration of justice in States.

Annex 13 does, however, express the importance of a balance in determining that fine line between the protection of safety and criminal accountability:

> National laws and regulations protecting safety information should ensure that a balance is struck between the need for the protection of safety information in order to improve aviation safety, and the need for the proper administration of justice.

It is a fine line of legal definition which has not been clarified in the Convention, for the excellent practical reason that it states: *Because of the different legal systems in States, the legal guidance must allow States the flexibility to draft their laws and regulations in accordance with their national policies and practices.*

It was this which demanded the genesis of a Just Culture, to promote continuous learning and encourage pilots to share essential information, which is the backbone of a safety culture, while protecting the individual's rights to fairness. Such a policy must, however, acknowledge the State's criminal process and thus ensure that the policy is fairly and constructively balanced with the accountability of the individual.

HOW DOES A JUST CULTURE POLICY SERVE ITS PURPOSE?

The European Organisation for the Safety of Air Navigation (Eurocontrol) is a pan-European organisation dedicated to supporting European aviation. It sees its priority in strengthening efforts to improve the safety of air operations over Europe, in particular with respect to incident reporting and data sharing, which it accepts is an area where progress is still slow. It euphemistically states that *one of the main elements towards progress* is the role and involvement of national justice authorities and the recognition and perhaps reconciliation of the responsibilities and interests of both the judiciary and safety authorities. It was this clash of authorities that led to Eurocontrol introducing the concept of Just Culture, as a means to guide those responsible in State domains for implementation and enforcement of legislation in the exercising their responsibilities within the concept of the Chicago Convention.

The Just Culture Task force is a relatively recent innovation, having been established by Eurocontrol in 2009 to promote debate and discussion on the legal issues that relate to safety and justice, to foster and support dialogue between safety and judicial experts and to develop relevant guidance material and policies in order to support the implementation and dissemination of Just Culture. The Task Force is composed of representatives from the Civil Aviation Authorities, Air Navigation Service Providers and Accident Investigators of the member States, the European

Commission and the European Aviation Safety Agency, a number of international air transport and air traffic control organisations (IATA, ERA, ECA and IFATCA), Accident Investigation Boards and, of course, State prosecution authorities.

The objective of the Just Culture Model Policy is to provide directions regarding the criminal investigation and prosecution of potential criminal offences resulting from incidents or so that both the safety investigation and judicial process can progress in parallel without either party acting in a manner prejudicial to the interests of the other. That being said, the whole process within the Model Policy places the objective of safety as paramount, ensuring a process that achieves a balance between the administration of justice and the safety requirements is essential and will maintain public confidence. To this extent, it relies on the foundation of established law:

- EU Regulation 996/2010 on the investigation and prevention of accidents and incidents in civil aviation;
- 376/2014 on the reporting, analysis and follow-up of occurrences in civil aviation;
- The Chicago Convention, Annex 13 (Accident and incident investigation) and Annex 19 (Safety Management).

The central requirement in all of these laws demands that accident and incident reports filed under an occurrence reporting scheme, as well as reports prepared or issued by investigating bodies or by the operators, but they should not be used as evidence in criminal proceedings against individuals. Prosecutors and other judicial authorities considering an investigation into an accident or incident should therefore take steps to gather evidence to support their own investigations rather than attempt to use that obtained by the safety investigation. Crucially, in addition to forensic or technical evidence, witness statements received in the context of a safety investigation are not to be used as evidence in criminal proceedings against the person who made it. If the State prosecuting authority wants to rely on witness evidence, it must be obtained independently of the safety investigation and in accordance with the Member States' criminal investigation and prosecuting procedures.

It must be borne in mind that the whole purpose of the technical investigation is to improve safety, not to assist criminal proceedings. However, there is nothing to prevent criminal investigations or evidence to be gathered under police and criminal evidence procedures, enabling them to interview witnesses as long as it is consistent with their statutory authority. Before the questions begin, the witness must be cautioned by the investigating officer, who must state:

> You do not have to say anything, but it may harm your defence if you do not mention, when questioned, something which you later rely on in court. Anything you do say may be given in evidence.

The witness is also entitled to legal representation and free and independent legal advice at a police interview under caution. The reader should be aware of the importance of ensuring that they have adequate advice and assistance, however; a solicitor experienced in criminal defence is essential, in order to advise their client and avoid the risk of self-incrimination when answering questions. Such representation may ensure that the witness does not say anything that can be misinterpreted or twisted by the police while they are questioning the witness or, indeed, under cross-examination at a Trial.

The Just Culture Task Force has a number of objectives which must be met in the way it is managed[8]:

- Promote debate and discussion on the legal issues that relate to safety and justice, in particular Just Culture;
- Foster and support dialogue between safety and judicial experts;
- Develop guidance material and policies in order to support the implementation and dissemination of Just Culture;
- Develop proposals for consideration by relevant bodies (Permanent Commission, European Commission / Ministries of Justice);
- Any other tasks considered necessary to meet the above objectives.

The International Civil Aviation Organization (ICAO) is a United Nations agency which helps States communicate and co-operate on aviation policies, with the intention of producing common understanding for their mutual benefit. In its 39th Assembly Session in 2016, a paper was presented by the Civil Air Navigation Services Organisation (CANSO),[9] an international organisation bringing together the air traffic management industry, on the topic of improving Just Culture. CANSO identifies the key feature of trust in a Just Culture as an atmosphere *in which people are encouraged for providing essential safety-related information, but in which they are also clear about where the line must be drawn between acceptable and unacceptable behaviour.* It sees a Just Culture as *one in which all employees are encouraged to provide, and feel comfortable providing, safety-related information. It is an environment in which employees understand they will be treated justly and fairly on the basis of their actions rather than the outcome of those actions, in the case of positive, as well as negative safety events. A Just Culture recognizes that systemic factors (not just individual actions) must be considered in the evaluation of safety performance and interpretation of human behaviour. A strong Just Culture in each aviation organization is perceived as the basis for a successful safety culture.*

Clearly, therefore, a Just Culture is seen as the foundation of the industry's essential safety culture, in which individuals at all levels do and say, when their commitment to safety is not necessarily under immediate scrutiny, knowing that they can place their trust in the Just Culture policy to protect them from the consequences of criminalisation.

What is essential is the part played by senior managers in putting in place an organised safety management system (SMS), implemented to ensure the safety of the aircraft or the service provider's management operation, detailing all the policies, practices and procedures that are to be followed in order to ensure the safe functioning of civil aviation in general and their part in particular. The SMS therefore ensures that every aircraft operation complies with the mandatory safety rules and regulations, and follows the codes, guidelines, and standards recommended by the ICAO, making it a core feature of the management and work practices of the organisation. Certainly, senior management commitment is crucial, but so, also, is the commitment of all the operating personnel, whether they be employees, partners or subcontractors, very possibly transcending national boundaries and jurisdictions; it is a safety system, not a legal system. That being said, the ICAO emphasises that Just Culture cannot and should not allow immunity from prosecution for deliberate or reckless acts of criminal negligence; but the question still remains: *Who gets to draw the line?*

While it is the company's obligation to define and document the responsibility so that it can be read and understood by all, it is the responsibility of accountable managers, including pilots and air traffic controllers, to ensure that the system is not just taken as a snapshot in time, but responds to events which manifest themselves in incidents, and thus demands a reporting culture that assures individuals of the same protection in Just Culture, maintaining that essential feature of trust. Consequently, the reporting system is crucial, particularly in the delivery of accurate and unvarnished safety reports from operating personnel themselves, which are viewed as some of the most valuable sources of information for learning safety lessons and identify safety trends. The policy therefore requires that constituent organisations foster a culture in which staff can take comfort that the organisation will treat them justly and fairly when they do report or when investigations and data analysis uncover safety risks. Most crucially of all, the reporting system demands that truth and accuracy must take priority over corporate loyalty, or the fear of consequences for disloyalty. Naturally, this is particularly relevant in circumstances in which personnel could feel caught in a dilemma involving internal disciplinary action or external criminal investigation as a result of actions which, yet, fall short of deliberate or reckless behaviour.

Given the theory and policy, a case study can illustrate its effectiveness.

Case Study: The Crash at Sachigo Lake

A North Star Basler BT-67 was on a flight from Red Lake to Sachigo Lake, Ontario, Canada, with two flight crew, was on approach to Sachigo Lake at about 09.10 on 3 December 2019, when the aircraft struck trees and crashed to the ground in a forest. There were no injuries but the aircraft sustained substantial damage.

On 3 December 2019, the pilot in command and first officer were scheduled to conduct a daytime cargo flight on board the DC3-TP67, on a domestic flight plan. During flight planning, the pilot checked the weather at 07.00 with forecasts for defined points in the flight. They duly departed under visual flight rules; but shortly after take-off, the aircraft entered, and climbed above, the cloud layers before reaching the planned cruising altitude, which was not in accordance with the applicable regulations for VFR flights. Before commencing their descent, the flight crew obtained the 08.00 hourly weather report, which had remained generally unchanged from the earlier report, and elected to carry out a visual approach to Runway 10. The pilot in command initiated a descent through the cloud layers by reference to the flight instruments.

Once the aircraft broke out of cloud cover at very low level, the aircraft was not in a position to continue with the planned visual approach. The pilot in command made low-level manoeuvres in an attempt to land, flying a large 360 degree turn, as low as

100 feet above ground level, some 400 feet below the required minimum altitude), and then flew a manoeuvre similar to a left-hand circuit, which brought the aircraft within close proximity to a 150-foot tall tower, in weather conditions below the VFR minimum requirements. The pilot in command had not briefed the co-pilot, who was unaware of his intentions and began calling out airspeed and altitudes. On the last attempt, during the low-level downwind leg, when the aircraft passed abeam the threshold of Runway 10, the pilot initiated a left-hand turn and began descending. About ten seconds later, the aircraft collided with terrain, in a near wings-level attitude, approximately 650 feet southwest of the threshold of Runway 10. The aircraft slid 350 feet southwards along the ground before it came to a rest on a south-westerly heading.

The investigation took some time but, on 4 August 2022, the Canadian Transportation Safety Board (CTSB) released their final report. They concluded that the probable causes involved a number of factors arising out of pilot errors that are commonly associated with fatigue and complacency in meeting routinely heavy workloads. Such might not offer a defence in criminal proceedings but heavily mitigate culpability and, more importantly, address a golden thread running through aviation accidents that can be vital in finding solutions to the problem in the future, if only the shadow of criminalisation can be lifted to assist the process of aviation safety:

- The decision to depart on, and continue, a visual flight rules flight in instrument meteorological conditions was influenced by a distorted perception of risk resulting from successful past experience in similar situations.
- The captain likely experienced attentional narrowing while carrying out a high workload visual approach at very low altitude in instrument meteorological conditions. This most likely resulted in an inadvertent but controlled descent that was not detected until the aircraft collided with terrain.
- The result-oriented subculture of some North Star Air's DC3-TP67 pilots, which emphasized mission completion over regulatory compliance, resulted in visual flight rules flights, such as the occurrence flight, being conducted in instrument meteorological conditions.

The report concluded that the pilot in command likely experienced attentional narrowing while carrying out a high-workload visual approach at very low altitude in instrument meteorological conditions (IMC). This most likely resulted in an inadvertent but controlled descent that was not detected until the aircraft collided with terrain. The uninjured pilots evacuated the aircraft via the right-hand cockpit window. The aircraft sustained substantial damage; however, there was no post-impact fire.

The collection of evidence invariably is assisted by the flight recorders; in fact, the regulations did not require this aircraft to be equipped with a flight data recorder or a cockpit voice recorder, North Star Air had installed a cockpit voice recorder (CVR), but following repairs, a paperwork error resulted in the CVR being placed in inventory instead of being re-installed in the occurrence aircraft. Without it, important evidence was not available to assist the investigation, as it made it more difficult to assess crew resource management, standard operating procedure execution and effectiveness, and workload management, with downstream consequences on the value of the report.

The investigation revealed that the result-oriented subculture of some North Star Air's DC3-TP67 pilots, which emphasised mission completion over regulatory compliance, resulted in VFR flights, such as the occurrence flight, being conducted in IMC. The investigation also determined the captain had a history of conducting VFR flights in IMC. The decision to depart on, and continue, a VFR flight in IMC was influenced by a distorted perception of risk resulting from successful past experience in similar situations.

In addition, the company structure was such that there was no operational management presence and no day-to-day direct pilot supervision of DC3-TP67 flight operations and flights crews at the company's remote bases, nor is it required by regulation. The absence of direct supervision meant that company pilots had considerable latitude when it came to making operational weather-related decisions. Over time, a culture of non-compliance developed that went undetected by the company management team. If minimal supervision of flight crews occurs within a company's flight operation, there is a risk that previously identified unsafe or noncompliant practices will persist.

Prior to this casualty, in December 2017, Transport Canada (TC) informed North Star Air that it had received allegations related to company DC3-TP67 aircraft operated on multiple occasions under VFR in IMC. TC provided the information about the allegations to the company for its internal investigation. In response, the operations manager contacted all the DC3-TP67 captains verbally and via email reminding them of the requirement to abide by the regulations, but the company did not outline any additional measures to monitor North Star Air's DC3-TP67 operation for reoccurrence. Then, in February 2018, TC initiated a reactive process inspection that included three flight inspections, which were deemed uneventful.

North Star Air had an SMS, even though at the time of the allegations it was not required by regulations. An SMS hazard report was created and entered into North Star Air's SMS database. However, the SMS did not identify the underlying factors that led to the reported instances of company aircraft operating VFR flights into IMC and the company did not take any additional measures to monitor its DC3-TP67 operation to ensure flights were being conducted in accordance with regulations. Additionally, a process inspection by TC in November 2028 did not include a review of the 2017 allegations; as a result, previously identified unsafe practices persisted, and played a direct role in this occurrence.

TC's monitoring of SMSs came in for criticism, as the Transportation Safety Board found a factor in TC's approach to surveillance resulted in North Star Air's SMS being approved even though several elements were not fully implemented. The report stated:

> If TC approves a company's SMS without first conducting an in-depth review to ensure that all required elements are present and effective, SMS enterprises may not have the ability to effectively manage safety. The TSB previously recommended that TC conduct regular SMS assessments to evaluate the capability of operators to effectively manage safety.

The Transportation Safety Board had implemented the Just Culture policy by identifying the causes, and making recommendations for the improvement of air safety, without the need for a threat of prosecution by the judicial authorities. As it subsequently reported in general terms:

> Following the occurrence, North Star Air implemented a flight operations quality assurance program. Additionally, in December 2020, TC conducted a PI focused on the evaluation and effectiveness of the long-term corrective action plan related to the flight operations findings from the December 2019 PI. TC concluded that the long-term corrective actions taken by North Star Air were effective.[10]

In this case, the CTSB concluded that a series of operational decisions during the flight resulted in increased risk levels, which were not isolated incidents but, over time, the pilot's repeated success conducting these types of operations resulted in a reduced perception of risk This contributed to the pilot's willingness to deviate from regulations, despite the well-documented risks associated with operating under visual flight rules in reported IMC.

TC had approved North Star's SMS, but the report cast doubt on the Board's detailed review to ensure that all required elements were present and effective, concluding that North Star Air's SMS did not identify the underlying factors that led to the reported instances of company aircraft operating visual flight rules flights into IMC, nor were any additional measures taken to monitor its DC3-TP67 operation to ensure flights were being conducted in accordance with regulations. As a result, previously identified unsafe practices persisted, and played a direct role in this occurrence.

It is clear, therefore, that expert investigations into the circumstances of an accident are crucial in a case, but a great deal of mitigation can follow, increasing the importance of a Just Culture that is intended to guarantee the process of investigation while minimising fear of prosecution. In any event, the prosecuting authority will have to make a decision based on the evidence before them, before they will consider whether an individual (or a company) should properly face criminal accountability. In this case, the value of a Just Culture was justified, without recourse to prosecution. Following the accident report, the company implemented a radical quality-control management plan, and after a subsequent audit in 2020, the TSB reported a satisfactory outcome. The company's president and chief operating officer had responded to the report with immediate changes to procedural and personnel changes within the operation of their Basler aircraft, including filing paperwork correctly, and observance of safety procedures.

THE POLICY AND THE INDIVIDUAL'S RIGHTS TO FAIRNESS

Clearly the starting point is to define the individual's concept of fairness in the context of their accountability for the consequences of an incident. Clearly, the institution in which people operate, be it an airline, air traffic control or aircraft service provider, gives the context which determines what people are entitled to expect in their rights and responsibilities, rather than some vague moral characteristic. After all, people participate in the rules and structures of the institutions in which they function. These rules determine the purpose of the institution, and reliance on such rules defines the responsibility and accountability of the individual. In the discharge of their responsibility, they have legitimate expectations on their rights and obligations.

Fairness has to be harnessed to responsibility in the workplace and, although guided by all manner of practical regulations and guidance, it must work within just four parameters, which have been defined in multiple sources:

- All management systems can be improved, by learning from experience;
- Human beings make mistakes;
- People can perpetuate errors for all sorts of systemic reasons, which can lead to unsafe patterns of behaviour in the work culture; and
- Reckless or wilful misconduct will not be tolerated.

These parameters are not defined by answers, but each is defined by questions, which individuals must ask themselves in any given situation. Upon this foundation, an individual will reflect on what is fair to them, which leads to the next level of reconciling Just Culture with the individual's rights to fairness. In the very practical environment of the workplace, Just Culture balances the need to have a non-punitive reporting and continuous learning environment, in which individuals must accept responsibility and, thus accountability, according to their obligations.

Naturally, the behaviour of an individual may fall below expectations such is the human element – but they may not always be in the wrong as there may be underlying work culture factors or system deficiencies affecting individual or group decision-making. Because civil aviation is so safety-critical, Just Culture recognises that there will be situations where safety mismanagement, at individual or group level, will fall below expectation. As a result, Just Culture delivers a managed process for dealing with such issues transparently and equitably.

The managed process goes further, and strives to improve the overall organisational culture and the safety performance of the organisation, not only by implementing new systems but also by encouraging the individual to take responsibility and ownership in their actions.

It is how the individual reflects on the answers to those questions, that shapes their concept of fairness, and leads to three golden features which characterise the system:

- Just Culture should focus attention predominately at identifying and addressing the constituent factors that deliver or inhibit risk management.
- It must provide an atmosphere of trust in which people are encouraged to engage in the process of providing and improving risk management, such as identifying, assessing and treating risks, within a non-punitive reporting culture.
- It must define unambiguously where the line is drawn between acceptable and unacceptable conduct, so that they individual is fully confident of where they stand.

Civil aviation demands that everybody in the operational process take responsibility for acting safely and in a manner which is consistent with their training, experience, and the professional standards expected of them. At the very least, the evidence trail demands that they meet written procedures unless, in the clear interest of safety, it is necessary to deviate from these procedures. Where such deviation is required, the individuals concerned will be given full and fair opportunity to account for their actions. Upon this foundation, the purpose of Just Culture demands honest reporting and discussing safety issues and mistakes, without fear of some punitive action, save the situation involving deliberate or reckless behaviour which, in any event, would be open to police and criminal evidence procedures.

The trust between the organisation and the staff in safety management is of paramount importance, and the guidance must be both unambiguous and supportive. Such risk and safety management procedures forming this foundation must therefore be robust in order to support the Just Culture procedures, which must be defined in a policy statement accompanied by a set of procedures describing the mechanism of how the Just Culture policy will be implemented, with fairness as a very visible priority. For the purpose of ensuring that the policy is understood and therefore followed, there must be some code of conduct to include items such as:

- How safety reports will be kept under restricted access when necessary, preserving confidentiality?
- Who will manage the safety reports?
- How the reporter and reported will be protected from adverse consequences and supported in the event of external litigation or threats?
- How the human element will be incorporated into the report findings, such as issues of fatigue and complacency in the human response?

The mischief of criminalisation, which justifies the purpose of Just Culture, has deep roots in unfairness, thus the policy must redress the imbalance of a judicial process which prioritises the demands of a risk society with criminal sanctions over the need to improve safety. The CAA has gone some way to rationalise this without antagonising the prosecuting authority, describing the aim of Just Culture *to promote continuous learning from previous mistakes and to encourage pilots to openly and freely share essential safety related information.* It states its approach as one in which people are treated fairly and are not punished for actions, omissions or decisions taken by them that are consistent with their experience and training. In other words, a Just Culture is intended to protect the individual's human rights and thus prevent a case of criminalisation arising in a case which previously had not been considered to be a crime. However, the CAA then qualifies its position that, in order to maintain or improve aviation safety in cases of, for example, gross negligence, wilful and reckless violations, further action may be necessary.

The CAA emphasises the need for the civil aviation community to feel they can openly report any incidents to the CAA without fear of punishment and the information they provide will be used for safety mitigation purposes and guidance only. They then state that *Where a pilot is found negligent that regulatory action may be deemed appropriate.* It is unclear how they define 'negligent,' which is a civil tort and open to challenges of definition and application; negligence itself is not a crime, but, globally, criminal negligence has been applied in civil aviation cases that undermine this theory of Just Culture, and the provisions of guidance in current law do not assist, as seen in the Civil Aviation (Investigation of Air Accidents and Incidents) Regulations 2018 which states

> 8. The sole objective of a safety investigation which is undertaken pursuant to… these Regulations is the prevention of accidents and incidents, without the apportionment of blame or liability.

We may summarise the purpose of Just Culture, therefore, to strike the right balance between the need for improving aviation safety and the sovereign right of the State's criminal jurisdiction process to investigate and prosecute a crime. The process necessarily involves a potential clash between the dispassionate technical investigation with a view to avoiding accidents in future, and the normative ethics of Society which expresses emotional demands that meet the criminal process. The theory is based on the need for a willing but independent partnership between the investigators and the judiciary, but the partnership becomes fractious if one partner believes that the other cannot be trusted, as we have seen in the case of the Mulhouse investigation which we have discussed.

But we are left bewildered as to how the findings of an investigation can be used against a witness who complies with the process. Drawing together the two strands of criminalisation and a Just Culture, we need to acknowledge the gorilla in the corner, and the consequence that the process of criminalisation has overridden the policy of Just Culture and, thus, undermined the confidence of the individual in the fairness of the system.

The separation of powers is an ancient concept, which democratic systems strive to achieve (and sometimes succeed), that the principal institutions of State, the Legislature (Parliament), Executive (the Civil Service) and Judiciary (the Courts) should be divided in body and function in order to safeguard liberties and guard against tyranny, the very pillars of State fairness. In practice, this means that Parliament implements laws, tells the Executive to jump, and the Courts must dispense justice by applying the Will of Parliament.

So, for example, Parliament enacts laws, which the CAA, through the Department for Transport must implement, and then prosecutions will be managed in the Court process, which will not be influenced by anybody. Aviation safety legislation applicable in the UK includes Statutes passed by Parliament such as the Aviation Security Act 1982, Statutory Instruments written by the CAA such as the Air Navigation Orders, and European Regulations which create criminal offences. We have met the Investigations and Enforcement Team of the CAA, which investigates potential violations of the criminal law that give rise to serious risks of air safety.[11] The case is then referred to the Courts, which are entirely independent of the other two institutions. It is worth restating that the most serious cases are triable only on indictment, in the Crown Court, when a Jury will decide on Guilt or otherwise according to the evidence before them, and if found Guilty the Judge will sentence the Defendant. Less serious cases are tried on an either way basis, meaning that the Magistrates decide whether a case is sufficiently serious to be heard in the Crown Court, but if they decide that the case should stay in the magistrates' court, the Defendant can then choose whether their trial should take place in the Crown Court instead. The least serious cases are tried summarily in the Magistrates Court.

The key issue is that, the moment the case is referred for prosecution, the Just Culture policy no longer plays a part. The individual's rights to fairness are a matter for State legislation on human rights:

> Article 5: Everyone has the right to liberty and security of person. No one shall be deprived of his liberty save in the following cases and in accordance with a procedure prescribed by law.
> Article 6: Everyone is entitled to a fair and public hearing within a reasonable time by an independent and impartial tribunal established by law.
> Article 7: No one shall be held guilty of any criminal offence on account of any act or omission which did not constitute a criminal offence under national or international law at the time when it was committed.[12]

That being said, these rights do have some application to Just Culture, because the management of the policy being applied must accord rights of fairness to the individual. In consequence, if an allegation is made in the Just Culture process, the individual must be entitled to notice of each allegation against them and must have the right to be heard and answer each such allegation.

HOW IS THE POLICY COMPROMISED BY THE CRIMINALISATION PROCESS?

In Chapter 2, we were introduced to the Risk Society, which has become a dominant feature in twenty first century social culture, demanding that somebody must take the blame for wrongs which the risk society perceives it to have suffered, and there must be some causal link between the wrong and the responsibility of the pilot or other professional, who must accept accountability even in the absence of the characteristics of a crime. The blame culture identifies this as beneficial, taking the consequences of risk management as the underlying justification to restore a balance dictated by society's conscience.

A central feature of this view is the human response to a situation, which falls below the threshold of criminal accountability, but all too often is identified as the cause of an incident, when, in fact, human error is an indicator of a deeper problem in the system, and it was this deeper problem which was the root cause; but the individual will still be prosecuted, because that is what the blame culture demands. When stripped bare of legal argument, this is the underlying factor that characterises criminalisation, giving rise to the Just Culture policy as a means to defend the innocent from the mischief, in order to meet the aims of reducing the risk of accidents in future; to this extent, the protection of the individual's human rights is a beneficial byproduct.

It is the human response which defines the objective of the blame culture, and therefore characterises the nature and purpose of the judicial investigation, to determine accountability for any behaviour that violates rules in law; in a Just Culture, accountability is defined by considering the physical, social and organisational context in which errors and violations took place, putting the error into context with the aim of resolving the problems revealed by the symptoms. If we apply this to the individual, the Just Culture will hold people accountable for the purpose of contributing to the constant learning process of safety management, not for the prosecution of offences, which is intended to prove guilt beyond reasonable doubt on the evidence.

Case Study: Mount Erebus 1979

Flight 901 routinely left Auckland International Airport at 08.00 for Antarctica, and was scheduled to return to Christchurch International Airport at 19.00, a tourist flight of 5,360 miles (8,630 km), with a scheduled 45-minute stop at Christchurch for refuelling and a crew change, before flying the remaining 464 miles to Auckland, arriving at 21.00. The aircraft used on the Antarctic flights were Air New Zealand's eight McDonnell Douglas DC-10-30 three-engined jets. The aircraft on 28 November was the fourth of the type acquired by Air New Zealand, which accepted hand-over on 12 December 1974, and logged some 20,700 flight hours by the date of the crash.

The pilot in command was Captain Thomas Collins, an experienced pilot who had accumulated 11,151 flight hours, including 2,872 hours in the DC-10. His co-pilot, Gregory Cassin, had accumulated 7,934 flight hours, including 1,361 in the DC-10. Flight Engineer Gordon Brooks had 10,886 flight hours, including 3,000 in the DC-10. Two other Air New Zealand flight crew were also travelling as passengers.

Some three weeks before the incident, the two pilots attended a briefing in which they were given a copy of the flight plan, that had been approved in 1977 but, when subsequently computerised, a typing error put the flight on a more southerly path, which took the aircraft directly over Mount Erebus, rather than along its side; but the ground staff had not alerted the flight crew to the change. Unaware of the fatal error, they entered this flight plan's coordinates into their aircraft inertial navigational system. Visual observation of their position to the mountain was impossible because of whiteout conditions, but both pilots had confidence in the extreme accuracy of the flight plan in the computer. The transcript of the cockpit voice recorder indicated that they believed that they were flying over McMurdo Sound, well to the west of Mount Erebus, but in fact they were flying directly towards the mountain. Then, at 12.49, the ground proximity warning system began sounding a series of alarms, warning that the aircraft was too close to the terrain. The pilots had begun a terrain escape manoeuvre, but it was too late. All 237 passengers and 20 crew on board were killed.

Ronald Chippindale was the Chief Inspector of Air Accidents in charge of the New Zealand Office of Air Accidents Investigations and was responsible for the investigation report. In his findings, he cited pilot error as the principal cause of the accident:

> The probable cause of this accident was the decision of the captain to continue the flight at low level toward an area of poor surface and horizon definition when the crew was not certain of their position and the subsequent inability to detect the rising terrain which intercepted the aircraft's flight path.

Chippindale did not level blame for the defective co-ordinates contained in the flight plan that had been fed into the aircraft's computer. However, his report was challenged and, in the face of public pressure, the New Zealand government announced a Royal Commission of Inquiry into the accident, to be chaired by Justice Peter

Mahon; but the demand was for the production of a report in an absurdly short space of time, exerting pressure on Mahon to meet, but with which he had to comply. Nevertheless, his report,[13] published in April 1981, exonerated the crew flight crew from blame, and held

> the single dominant and effective cause of the disaster was the mistake made by those airline officials who programmed the aircraft to fly directly at Mt. Erebus and omitted to tell the aircrew. That mistake is directly attributable, not so much to the persons who made it, but to the incompetent administrative airline procedures which made the mistake possible.

He concluded:

> In my opinion, neither Captain Collins nor First Officer Cassin nor the flight engineers made any error which contributed to the disaster, and were not responsible for its occurrence.

The condemning factor in Mahon's Report found that airline executives and senior pilots had engaged in a conspiracy to whitewash the inquiry, accusing them of an orchestrated litany of lies by concealing relevant evidence and giving false statements to investigators. As for Chippindale's original report, Mahon found that he had a poor grasp of the flying involved in jet-airline operation, and that his investigation techniques lacked rigour, which allowed errors and avoidable gaps in knowledge to appear in reports. Consequently, Mahon held, Chippindale entirely missed the importance of the flight-plan change and the rare meteorological conditions of Antarctica. Had the pilots been informed of the flight plan change, the crash would have been avoided.

Within a month, Air New Zealand applied to the High Court of New Zealand for judicial review of a number of the findings in Mahon's report and against Mahon's order that it pay more than half the costs of the Mahon inquiry. The application was duly referred to the Court of Appeal on the basis that the Judge had exceeded his powers for a judicial review in making the order for costs, but the appeal went further than that, as it was alleged that the Judge's decision was flawed because of a combination of failures to meet the Rules of Natural Justice. These Rules were defined at the ultimate hearing as follows:

> The first rule, is that the decision to make the finding must be based upon some material that tends logically to show the existence of facts consistent with the finding and that the reasoning supportive of the finding, if it be disclosed, is not logically self-contradictory.
>
> The second rule requires that any person represented at the inquiry who will be adversely affected by the decision to make the finding should not be left in the dark as to the risk of the finding being made and thus deprived of any opportunity to adduce additional material of probative value which, had it been placed before the decision-maker, might have deterred him from making the finding even though it cannot be predicated that it would inevitably have had that result.

The difficulty encountered with Mahon's conclusions was that they did not all lead to findings as to the cause of the crash but that members of the management of Air New Zealand had conspired to commit perjury before the inquiry *afterwards*, to cover up the errors of the ground staff. In the event, the case was finally appealed to the Judicial Committee of the Privy Council in London (JCPC) the ultimate Court of Appeal, who concluded that Mahon did not have a discretionary power to make an order for costs and that *to this extent, if to no other, the Royal Commission Report was subject to review under* section 4 *of the Judicature Amendment Act.*

Their Lordships had a great deal more to say about the conduct of Air New Zealand executives in the investigation process. The three matters highlighted in the appeal were:

1. The deliberate destruction on the orders of Mr. Davis, the Chief Executive, of all documents which would disclose the mistake that had been made over the co-ordinates in the flight plan used for briefing Captain Collins and the different co-ordinates in the flight plan issued to the aircrew for use in the aircraft's computer on the actual flight.
2. The concealment that there had been an intentional adoption by the airline management, as the southernmost waypoint for their Antarctic sightseeing flights, of the waypoint used at Captain Collins' briefing with co-ordinates that would take the aircraft over ice-covered sea to the west of Mt. Erebus and well clear of it.
3. The denial by senior officials of ANZ that they knew that aircraft engaged on sightseeing flights to Antarctica, when they got there in visual meteorological conditions ("VMC") with visibility at 20 kilometres or more, flew at heights lower than 6,000 feet.

The JCPC approved much of Mahon's report that dwelt on the incompetence – or worse – of Air New Zealand, which provides a cautionary tale for operators in the future:

> The JCPC Judge's Report contains numerous examples and criticisms of ANZ's slipshod system of administration and absence of liaison both between sections and between individual members of sections in the branch of management that was concerned with flight operations. Grave deficiencies are exposed in the briefing for Antarctic flights; and the explanation advanced by witnesses for the airline as to how it came about that Captain Collins and First Officer Cassin were briefed on a flight path that took the aircraft over the ice-covered waters of McMurdo Sound well to the west of Mt. Erebus but were issued for use in the aircraft's computer as the nav track a flight path which went directly over Mt. Erebus itself, without the aircrew being told of the change, involved admissions of a whole succession of inexcusable blunders by individual members of the executive staff. None of this was challenged before their Lordships. No attempt was made on behalf of ANZ to advance excuses for it.

Air New Zealand did not attempt to challenge the findings of dishonesty, because they did not need to. The grievance against Mahon was that he had exceeded his powers under the Rules of Natural Justice; allegations of dishonesty and gross negligence were matters for other courts, if prosecuting authorities so-decided. Lord Diplock concluded in the JCPC Judgment:

> These appalling blunders and deficiencies, the existence of which emerged piecemeal in the course of the 75 days of hearings, had caused the loss of

257 lives. Their Lordships can well understand the growing indignation of the Judge when, after completing the hearings and for the purpose of preparing his Report, he brought them together in his own mind and reflected upon them. In relation to the three matters that were principally canvassed in this appeal and upon which he based his finding that there had been a pre-determined plan to deceive the Royal Commission and a conspiracy to commit perjury at its hearings, their Lordships have very reluctantly felt compelled to hold that, in the various respects to which their Lordships have referred, the Judge failed to adhere to those rules of natural justice that are appropriate to an inquiry of the kind that he was conducting and that in consequence it was not open to him to make the finding that he did in paragraph 377 of his report.[14]

SUMMING UP

An enormous amount of work has been published on the Erebus case and the issues of the conflicting reports, but the thread running through this case demonstrates the threat of criminalisation in air accidents. If Mahon were correct in his findings of a pre-determined plan to commit perjury, then it would be open to the New Zealand prosecuting authorities to deal with that. The blame culture which fuelled the national outrage should not have infected public opinion about the outcome of the technical investigation, but rather the value of Chippindale's original report. This case predated the Mulhouse crash[15] by nine years, but the themes of distrust of non-legal investigations cannot be ignored. Ultimately, the normative ethics of society drive laws and procedures, so it is a matter for the judicial process to determine on prosecutions according to the policies driven by the normative ethics of society. If the technical investigation concludes that there was evidence of perjury, or that the technical investigation was itself tainted by dishonesty, then that is for the criminal investigation process to address in proceeding against individuals based on the evidence, rather than to take matters into their own hands, as the French prosecuting authorities did in the Mulhouse crash or Mahon did in the Erebus crash.

From the point of view of the individual, the key feature defining Just Culture is that of fairness, and the adoption of its policy is designed to assure the individual of that quality of fairness, in meeting the priority of investigating the facts to improve air safety. The cooperation of all the surviving individuals is essential in the investigation process; what may trouble individuals, however, is the fear of prosecution as a result of the evidence collected. Of course, there will be occasions when an individual may respond dishonestly in order to conceal their wrong-doing, which does not speak highly of human nature but is a fact of life. We must address the scenario, however, in which the individual is in fear of criminalisation, that is, their accountability for a crime that leads to their prosecution even though they were not considered crimes by act or omission, but which have been made crimes by Society notwithstanding the human rights of the individual. Even though occurrence reports should be used only for the purpose for which they were intended, the fear of self-incrimination (and, indeed, disclosure as we found in *Hoyle v Rogers*) may discourage such cooperation, defeating the purpose of accident reporting. There is also the issue arising that, if one party files a report, then, however the other feels about the incident, questions will be raised about the investigation process. In the case of the Darwin Airline Saab crash which we discussed in Chapter 2, the Swiss accident investigation was complicated by the fact that neither the crew of the Sportcruiser nor the tower controller submitted a report. The crew of the Sportcruiser argued that they were not in a position to assess the risk adequately and therefore did not file a report, while the tower assessed the situation as non-critical. Only the crew of the Saab submitted a report and thus initiated the investigation into the occurrence. The investigators observed that just because other parties are involved in an incident does not release one from the obligation to submit a report.

Naturally, the fear of criminalisation becomes very real indeed when the investigation is conducted by law enforcement officials, or if the process is taken out of the hands of air accident investigators. Prosecuting authorities lacking specialist aviation experience may take a misconceived view of a pilot's risk assessment in a situation, which pilots must manage on a daily basis; accident investigators are well-acquainted with such scenarios, but law enforcement officials may only have the Code of Crown Prosecutors to guide them, and the risk assessment finds itself at the centre of a Trial, which a Jury must decide according to the law. If the professional obediently cooperates with an investigation, and then finds the report admitted in evidence, then the most that they can do is to apply to have it disallowed; but even the air accident investigators are worried in that case, because it is a maxim in law that the Defence must have the opportunity to cross-examine a witness in Court, certainly trying to discredit their evidence or findings, as a result of which the investigator might fear for their own reputation. The whole policy therefore is thrown into question.

This raises another issue, addressed by Professor Dekker, as to what and where the role of domain expertise should lie in judging whether or not the professional's risk assessment and behaviour is acceptable within the norms of their professional duty. Clearly, the reconstruction work in an accident investigation must be conducted by experts, thus avoiding hindsight bias, which would be a serious trap for a law enforcement officer. The risk management and response of the pilot can be contextualised much more effectively by a technical investigator who can assess the individual's behaviour in a crisis situation, in which they may have exercised their experience and training more effectively than a prosecutor would discover in any written regulations. As a result, we can conclude that, in a Just Culture policy, the line gets to be drawn by experts who understand the underlying norms in the job, and can draw their findings accordingly. If, having drawn their conclusions, the evidence against the professional demonstrates dangerous or reckless conduct giving rise to the *mens rea* and *actus reus* of an offence,[16] then the criminal legal process will follow, and the balance struck between the Just Culture and the Process of Law.

That being said, Society may have a say in a casualty which gains publicity and demand that the question be answered: *Who is to blame?* In that case, the prosecuting authority will be persuaded that it would be in the interests of justice to prosecute the individual, having satisfied themselves that the Full Code test has been satisfied, with evidence upon which a Jury could convict, and it would be in the public interest to do so. Innocent victims would also have the opportunity of claiming compensation in civil proceedings. This can have an important bearing on the rights of individual claimants. In civil proceedings, the Claimant must establish the Defendant's liability on the balance of probabilities; in other words, the evidence must determine that it was more likely than not that the Defendant was liable. In criminal proceedings, the burden is higher, in that the Prosecution must establish beyond reasonable doubt that the Defendant was guilty; as Judges sum up to the Jury, they must be sure. This has the effect of rendering a criminal conviction more certain in evidential terms, and Section 11 the Civil Evidence Act 1968 permits a claimant to adduce such evidence in their Statement of Claim:

> 1. In any civil proceedings the fact that a person has been convicted of an offence by or before any court in the United Kingdom… shall… be admissible in evidence for the purpose of proving, where to do so is relevant to any issue in those proceedings, that he committed that offence, whether he was so convicted upon a plea of guilty or otherwise and whether or not he is a party to the civil proceedings; but no conviction other than a subsisting one shall be admissible in evidence by virtue of this section.
> 2. In any civil proceedings in which by virtue of this section a person is proved to have been convicted of an offence by or before any court in the United Kingdom –
>   a. he shall be taken to have committed that offence unless the contrary is proved; and
>   b. without prejudice to the reception of any other admissible evidence for the purpose of identifying the facts on which the conviction was based, the contents of any document which is admissible as evidence of the conviction, and the contents of the information, complaint, indictment or charge-sheet on which the person in question was convicted, shall be admissible in evidence for that purpose.

In this case, the consequence does not have a negative effect on the Just Culture policy, as the balance of fairness has not been compromised, unless the conviction arose out of an abuse of the process in which the accident investigation led and the criminal process followed in its wake, exploiting evidence in which the Defendant incriminated themselves. It is the fear of the downstream consequence which troubles the professional, distorting the balance between the accident investigation and the criminal one. The theory of the Just Culture policy is to define and understand the line between acceptable and unacceptable behaviour, which the prosecuting authority will follow in the wake of the casualty report. The success or otherwise of the theory can be assessed from case studies, which we shall examine in the next chapter.

NOTES

1 *Andrews v DPP* [1937] AC 576 HL.
2 The conviction was upheld.
3 See Chapter 1.
4 Dekker S, 2012, *Just Culture: Balancing Safety and Accountability*, Taylor & Francis Group, Boca Raton, FI.
5 Condensed from The Code for Crown Prosecutors (currently 8th Edition, October 2018), https://cps.gov.uk.
6 Which is beyond reasonable doubt.

7 In limited circumstances, where the Full Code Test is not met, the Threshold Test may be applied to charge a suspect, but that is beyond the scope of this book.
8 Eurocontrol, 2012, Just Culture Policy.
9 ICAO A39-WP/193 TE/73. See www.icao.int/Meetings/a39/Documents/.
10 Transportation Safety Board of Canada, Air Transportation Safety Investigation Report A19C0145 (released 04 August 2022).
11 CAA, 2020, Code of Practice for the Investigations and Enforcement Team, CAP 1422.
12 See European Convention on Human Rights, scheduled to the UK Human Rights Act 1998 as just one of many jurisdictions.
13 A full analysis of the reports can be found: www.erebus.co.nz/Investigation/Mahon-Report. Thorough as it is, the author does not endorse opinions or conclusions expressed in this website.
14 *The Honourable Peter Thomas Mahon v Air New Zealand Limited and Others* from the Court of Appeal of New Zealand, Privy Council Appeal No. 12 of 1983.
15 See Chapter 3.
16 It will be recalled that evidence of *mens rea* is not required in strict liability cases, such as the breach of certain Regulations.

5 What Do Case Studies Tell Us about Just Culture and Accountability?

OVERVIEW

The gulf between professional behaviour and criminal liability rests, ultimately, on how the law determines blameworthiness. If the differences between the two are irreconcilable, then the gulf must be unbridgeable, which essentially defines the philosophy giving rise to a Just Culture policy. The trouble is that there are no clear lines of accountability in risk; ultimately, it is the prosecuting authority who must draw the line in any given case, and their perception of that line may be very fluid, and influenced by their knowledge and experience in the law.

The question of risk necessarily involves effective management of the risk:

The main elements of risk management form a key underlying feature in Just Culture, which defines the steps in assessing and treating the risk:

- classifying work activities
- identifying hazards and personnel at risk
- determining the likelihood of the risk event happening, against the consequences if it does happen;
- deciding if the risk is tolerable
- preparing an action plan (if necessary)
- reviewing the adequacy of the action plan

We have examined in detail how the Just Culture policy works in principle, to assess the parameters of performance. What is important now is to identify how Just Culture needs to work in practice, by applying the factors of risk management to promote continuous learning from previous mistakes and to encourage pilots and air traffic control to openly and freely share essential safety-related information.

That necessarily involves the application of a Just Culture in practice, in scenarios that may well be far from routine and take us into the realm of foreseeability.

Failure to manage the risk can lead to allegations of recklessness and gross negligence, giving rise to a management failure. This is where the Corporate Manslaughter and Corporate Homicide Act 2007 comes in with a new layer of criminalisation and draws in the issue of corporate accountability, which defines another facet of the policy that puts corporate players in the dock, who

DOI: 10.1201/9781003489696-5

are responsible for the management of an organisation that causes a person's death and amounts to a gross breach of a relevant duty of care owed by the organisation to the deceased. As such, the senior air traffic controller or any senior manager whose acts or omissions expose the organisation to have a substantial element in the breach will incriminate their employer.

THE BALANCE OF FAIRNESS

The keystone of a Just Culture is treating the professional fairly when errors occur leading to an incident. Fundamentally, it is a mindset within the professional environment that starts from an assumption that professionals set out to act in good faith, with honest intent to respond to a crisis situation for the most favourable outcome, applying their experience and training to their discretion in delivering that outcome. It therefore depends on an analysis of the human response, which the Just Culture sees as a flexible resource which adapts to each scenario, thus delivering resilience, which if applied in a scenario leading to errors with unintentional rule violations, represents a low level of blame. Safety recommendations may well follow, but the professional may not have crossed the line that leads to a prosecution. If the error resulted from gross negligence or recklessness, then the prosecuting authority may be persuaded that the response crossed that fluid line of accountability; the balance of fairness has still been struck. The problem arises when the Prosecutor draws the line unfairly; but the professional needs to understand *how* the Prosecutor draws the line, and that demands an understanding of the law defining criminal negligence, if an unsafe decision is to be challenged in defence evidence in a given case. Once again, we are back with Chapter 1.[1]

The trap facing investigators lay, primarily, in how the balance of fairness is accomplished; all too often the trap lay in hindsight bias. What seemed like a sensible professional decision in the heat of a crisis situation may seem to be grossly negligent, or even reckless, in hindsight. The solution is to examine the facts from the individual's viewpoint by rational assessment, and that demands an understanding of their actions in context. In other words, the risk management function must be ascertained in the case, and that is more likely to be achieved fairly by technical accident investigators than by a Prosecution lawyer.

The process of risk management in aviation involves co-ordinates recognition of risk, risk assessment, developing strategies to manage it and mitigation of risk; it will be noted that the elimination of risk may be impossible, but it must be reduced to a level below the threshold for criminal accountability. Risk management will be specific in each scenario, but there are some general principles which can guide the assessment, and which will be generated organisationally, finding itself at the centre of the system in which the aircraft operates. These principles follow steps which will be found in every general risk management manual, in one form or another:

- Step 1: Identify the risk in context.
- Step 2: Map out the scope of the risk, to lay the foundation to answer the question: Is it an acceptable risk?

- Step 3: Consider the objectives of stakeholders, such as crew safety, the owners, insurers, operators and passengers – and people or property on the ground.
- Step 4: Define a framework and an agenda for the management of the risk.
- Step 5: Develop an analysis of the risks involved in the process.
- Step 6: Mitigation of risks using available technological, human and organisational resources.

With the general principles in mind, the professional will then apply them to the crisis situation, in order to decide which of the following treatment options should be adopted:

- Tolerate

  In other words, Do Nothing. Having balanced the issues, in particular the objectives to be achieved, the pilot, the air traffic controller, or some other professional may decide that the likelihood of the undesired event is remote by comparison with the consequence, bringing the risk within the tolerance of acceptability and requiring no action at all. These instances are likely to involve a low risk assessment. However, an unsafe decision resulting in disaster may be viewed as recklessness and may give rise to liability in negligence, or criminal accountability.
- Treat

  Treatment involves reducing the risk, in other words taking steps to mitigate or even minimise the consequences of the risk.
- Terminate

  If the consequences outweigh the likelihood, then the risk is deemed too high, and the risk should be terminated by avoiding it; for example, the risk following a technical alert or dangerous weather would provide grounds for interrupting or terminating the flight.
- Transfer

  The alternative is to transfer the risk to another party; which, after all, is the purpose of insurance, and a commercial analysis will often be a determining factor in the successful outcome of the flight. The danger lies in transferring a risk to which the pilot miscalculates that vital equation of balancing the likelihood of an event happening against the consequence if it does happen. Invariably the general risk management process will demand the utmost caution.

When reading these words calmly in black and white, we distort the truth that, within the critical space of moments, an action plan must be established, at least in the mind of the pilot or the air traffic controller, and then monitored and reviewed, until the crisis has been safely passed. This is another facet of the human element, demanding an understanding of the tensions in split-second decision-making, which utterly opposes the mischief of hindsight bias.

The risk management process spans many different disciplines in the aviation environment, particularly routing and air traffic control, and ground maintenance operations. In the Überlingen crash,[2] the air traffic controller had been overworked and was covering multiple tasks without support, and without essential functioning equipment. With the benefit of hindsight, questions may be addressed, such as why the controller would allow aircraft to fly on collision courses, or agree to single-controller operation in violation of the rules, but the examination demanded a real-time approach, rather than a hindsight one. In the Tuninter Flight 1153 crash,[3] technicians inadvertently installed the wrong fuel quantity indicator, which the pilot in command could not have foreseen; but he subsequently flew without evidence of a refuelling slip, in violation of the rules; and then panicked in the critical risk management process.

These issues raise questions about Just Culture and accountability that clearly can be examined more closely in case studies, in order to get a better understanding of the dynamics of risk management in Just Culture, which allows us to rationalise the principles of accountability.

Risk management goes further than instant analysis and response in a crisis; indeed it reaches into the heart of the crisis scenario, demanding foresight and anticipation of risks, which, after all, is the core purpose of learning from aviation incidents. Naturally, the balance of fairness must address reasonable foreseeability, which drills deeply into the tort of negligence, which we have seen has evolved from the principles in the case of *Donoghue v Stevenson*[4] to criminal accountability, as seen in *Adomako*.[5] Indeed, the Australian government's review panel recommended that all actions for personal injury and death resulting from negligence should be subject to a single legal regime regardless of whether they are brought in contract, tort, under a statute or under any other cause of action.[6] It recommended a legislative restatement of the current law to the effect that *in cases involving an allegation of negligence on the part of a person holding himself or herself out as possessing a particular skill, the standard of reasonable care should be determined by reference to what could reasonably be expected of a person professing that skill, judged at the date of the alleged negligence.*

The fundamental principles of liability have been stated before:

1. that the Defendant had a duty to avoid acts or omissions which might reasonably foreseeably cause the victim loss or damage;
2. that the Defendant was in breach of that duty of care;
3. that the Claimant suffered damage as a reasonably foreseeable consequence of the Defendant's breach of duty – in other words, the Defendant's breach of duty was the proximate cause of the damage; and
4. that taking the leap from civil to criminal accountability, the breach was so grossly negligent that it should be seen as criminal.

The balance of fairness demands scrutiny, on the one hand, of the professional's rights to fairness, ultimately defined in the European Convention on Human Rights, against, on the other hand, the legal rights demanded by Society in fairness to the common good, which Judges must divine from guidance on what the lawmakers intended when making the applicable Statute law, and precedent cases. The balance must then be applied to the facts of the individual case; hence, case studies prove essential in determining fairness.

Mistaken identity and hostile fire over enemy territory are among the many risks that pilots face when crossing war zones; in the twenty-first century it has become an occupational hazard. Two particular case studies enable us to analyse the consequences of risk in such conditions, and lead to a third which, when combined, cast doubt on fairness and transparency when it is applied to State actors, as distinct from professionals simply trying to do their jobs; but those professionals must learn lessons from the facts in the cases, in discharging their duties to the professional standards expected in air safety.

Case Study: Iran Air Flight 655

The back-story involves the Iran-Iraq war, which had become particularly dangerous to third parties in 1984, when Iran expanded military operations to include air attacks against oil tankers and merchant shipping of neighbouring countries, some of whom were providing aid to Iraq by shipping Iraqi oil. In May 1987, the Iraqi Air Force had attacked the American-guided missile frigate *USS Stark*, killing 37 American crew. In late 1987, American naval forces were exchanging gunfire with Iranian gunboats, and the guided missile frigate *USS Samuel B. Roberts* had struck an Iranian sea mine in April 1988. Two months before the incident, the United States took part in Operation Praying Mantis, resulting in the sinkings of the Iranian frigate *Sahand*, the Iranian fast attack craft *Joshan* and three Iranian speedboats. Additionally, the Iranian frigate *Sabalan* was severely damaged, two Iranian platforms were destroyed, and an Iranian fighter was damaged, with the loss of at least 56 Iranian crew were killed.

Tensions were thus running extremely high in the region, meaning that civilian air operators and routers had to make serious risk management decisions on flying in the region. On 3 July 1988, Iran Air Flight 655, a scheduled passenger flight from Tehran to Dubai via Bandar Abbas, took off from a joint civil / military airport and flew via commercial airway 59. The aircraft was an Airbus A300 which had been built in 1982. Earlier that morning, American naval forces monitored an attack by Iranian gunboats on a merchant ship, when an Iranian gunboat opened fire on a helicopter from *USS Vincennes*, a guided-missile cruiser. However, in the fog of war, Flight 655 was misidentified and shot down by two surface-to-air missiles fired by the *Vincennes*. The missiles hit the civilian aircraft while it was flying on the established commercial airway over Iran's territorial waters in the Persian Gulf, shortly after the flight departed its stopover location, Bandar Abbas International Airport. All 290 people on board were killed.

By any stretch of the imagination, the casualty was reasonably foreseeable, as the Iran-Iraq War had been continuing for nearly eight years, and the aircraft was in a high-risk area. *Vincennes* had entered Iranian territorial waters after one of its helicopters drew warning fire from Iranian speedboats operating within Iranian territorial limits. The *Vincennes* crew had incorrectly identified the Airbus as an attacking F-14 Tomcat, a fighter that had been part of the Iranian Air Force inventory since the 1970s, and which the crew had been warned had been equipped with air-to-ground missiles. The American president of the day, Ronald Regan, subsequently issued a written diplomatic note to the Iranian government, expressing deep regret, and later agreed that it should be considered an apology, although the American government continued to plead that the warship had acted in self-defence.

In a subsequent interview in 1989, Professor Andreas Lowenfeld from the *American Journal of International Law* criticised the official American position that they were not liable in law for the incident,[7] asserting that, under American law, liability is established regardless of fault, so long as the cause is proved, on the basis that responsibility flows from the action itself. In the light of our examination of the law, this raises the contentious issue that the proximate cause must give rise to liability, notwithstanding the absence of reasonable foreseeability. But the issue of reasonable foreseeability was still apparent in this case; the key underlying factor is whether the mistaken identification of the aircraft as a hostile fighter amounted to an error of judgment, or even blameless inadvertence. The question remains, whether the action taken by the American commander amounted to the breach of a duty of care. The case was never concluded in law, however, as it was settled in February 1996, when the United States paid compensation, without admission of liability.

Notwithstanding the actual facts of the shooting-down of the aircraft, the balance of fairness would inevitably demand answers to key questions, that can be asked in all such cases by airline pilots, air traffic controllers and routing managers:

- Was the routing of the aircraft made with a full risk assessment?
- Who conducted the risk assessment of the routing of the aircraft?
- Should the operator have sought re-routing?
- Should the pilot or dispatcher have sought re-routing?
- Should air traffic control have sought re-routing?

Airway routing occurs along pre-defined flight paths, which are three-dimensional highways; even in hazardous areas, the risk may be mitigated by flying at a higher altitude, reducing the risk of attack by surface-to-air missiles; naturally the alternative of a deviated route offers a safer option, but the cost in terms of operational expense can be vast. The administration authority for the State in question is responsible for providing safe and efficient air navigation services in the national airspace system, but the flight plan itself is filed with the administration authority by the pilot or the airline's flight dispatcher prior to departure, which indicates the aircraft's planned route. It therefore follows that a number of parties have influence or control over the system. Most airlines have teams tasked with analysing potential routes for demand and profitability; safety is a matter which sometimes takes a rear seat, which sometimes follows in second place to commercial priorities.

A report published by ICAO in 1988[8] particularly highlighted issues which had to be learned from the incident, which involved problems arising out of the presence and activities naval forces in the Gulf area, quoting air traffic controllers who had overheard challenges to civil aircraft on the military air distress frequency and were duly able to alert civil pilot. It pointed out that, in some cases, flights chose to re-route in order to avoid possible danger of attack, but thus *accepting a significant mileage penalty with its economic consequences and inconvenience to passengers.* In this instance, the report noted that, three minutes after take-off, seven challenges were made by USS *Vincennes* on the military air distress frequency, but civil aircraft did not carry radios that could tune to that frequency. Three more challenges were made by the *Vincennes* and one by another American warship on the international air distress frequency but all were unacknowledged. The report found that there was no indication of failure during flight in the equipment of the aircraft including the communications and navigation equipment. ICAO concluded with eight safety recommendations, all related to communication of information with military units. Among them, it was proposed that military units should be equipped to monitor appropriate air traffic control frequencies in order to enable them to identify radar contacts without communication, while pilots of civil aircraft should ensure the continuous operation of weather radars and radio altimeters, so that military pilots could pick up the electronic emissions.

The balance of fairness clearly put the obligation on the military assailant in this case. Twenty-six years later, another aircraft was brought down in a military attack, in a case which demonstrated how the assailant State can obfuscate facts with a vexatious defence that is designed to obstruct the whole process of fairness and, indeed, justice.

Case Study: Flight MH17

The back-story to this event is complicated.[9] In November 2013, Ukraine's new President Yanukovych suspended trade and association talks with the European Union, opting to revive economic ties with Russia. Protests began in Kyiv's Independence Square almost at once and, in the following month, President Yanukovych met Russian President Putin for talks, to lay the ground for a new *strategic partnership* between the two countries. Over the next three months, the civil unrest in Ukraine grew very much worse, and in February, Ukraine's parliament voted to impeach President Yanukovych, who promptly fled the country. Within days, dozens of pro-Russia gunmen seized the government buildings in the Crimea, sovereign territory of Ukraine, and raised the Russian flag. On 1 March, Russian Parliament approved President Putin's request for Russian forces to be used in Ukraine and, three weeks later, he signed a law formalising Russia's takeover of Crimea from Ukraine. By the end of March, Russia was believed to have amassed a force of several thousand troops close to Ukraine's eastern border. Then pro-Russian separatists seized regional government buildings in the Ukrainian cities of Donetsk, Luhansk and Kharkiv, calling for a referendum on independence by 11 May. It was obviously engineered by Moscow, and a ruse to endeavour to sidestep United Nations provisions protecting the sovereign powers of a nation, but the West did nothing except trade sharp words. On 14 June, pro-Russia separatists shot down a military transport plane in eastern Ukraine, killing all 49 Ukrainian service personnel on board. Indeed, intelligence reports found that 16 military aircraft had been shot down in eastern Ukraine in the weeks before the MH17 incident.

It has been argued, with some force, that Ukraine should have responded to this in meeting its duty of care and closed its airspace in eastern Ukraine to civilian air traffic. The route flown by MH17 was the most common route for flights from Europe to South East Asia, but some airlines had already treated the risk and re-routed their aircraft to avoid the area, accepting the higher costs as a consequence. British Airways and Air France had already altered their flight paths to avoid Ukraine. Many more airlines would follow after the event.

On 17 July 2014, Malaysia Airlines Flight 17 was scheduled to fly from Amsterdam Schiphol at mid-day, and took off at 12.31 with 298 passengers and crew, bound for Kuala Lumpur and due to arrive at 06:10 the next day. The flight plan was due to take the aircraft over Ukraine at a height of 33,000 feet before climbing to 35,000 feet north-east of Dnipropetrovsk. Accordingly, when it reached the area, Dnipropetrovsk Air Traffic Control requested that it climb to 35,000 as planned, and to maintain separation from another flight, Singapore Airlines Flight 351, which was also flying at 33,000 feet. The flight crew of MH17 asked to remain at 33,000 feet and the air traffic controller approved, moving the other flight to 35,0000 feet. At 16:00 local time, the crew asked for a deviation of 20 nautical miles to the left of the course, due to weather conditions. This request was approved by Dnipro Air Traffic Control. The crew then asked if they could climb to 34,000 feet, but this was refused as it was not available, so they remained at 33,000 feet. Some 20 minutes later, Dnipro noticed that the flight was 3.6 nautical miles north of where they should be, and instructed them to get back on course. Dnipro then telephoned Rostov-on-Don Air Traffic Control in Russia and requested clearance to transfer the flight to Russian airspace. After obtaining permission, Dnipro attempted to contact MH17 for handing them over to Rostov-on-Don, but the aircraft did not respond.

In the ensuing search of the wreckage of the downed airliner, the black box was found and the contents had survived for examination. A 2.3-millisecond noise was recorded on the cockpit's voice recorder before the system failed. By triangulating the signal, experts were able to show that it originated outside the aircraft. The Boeing 777 had been shot down by a missile using a warhead that detonated one yard outside and above the left-hand side of the cockpit, causing catastrophic damage to the forward part of the fuselage. The forensic evidence was collected and analysed in minute detail, which led to the confident conclusion that the detonation caused an in-flight break up that saw the forward part fall away while the after part of the fuselage continued in the air some five miles before disintegrating.

The investigators from the Dutch Safety Board found that the weapon used was a model warhead carried on a missile from the Buk surface-to-air missile system. From the damage pattern found on the wreckage caused by the blast and the impact of high-energy fragments, analysis of the explosive residues and the distinctive shape of some of the fragments, the evidence proved beyond reasonable doubt that the warhead was carried on a class of missile installed on the Russian-made Buk surface-to-air missile system. In drawing its conclusions, the board also found:

- The number of perforations in the fuselage was not consistent with air-to-air gunfire and none of the air-to-air missiles in use in the region had the distinctively shaped fragments.
- The type of fragments found in the wreckage and bodies of crew members in the cockpit were bow-tie or cube shaped – only found in the 9N314M warhead.

In a denial of liability bordering on the childish, the Russian authorities disputed the Dutch findings, insisting that only older missile systems, of the type still in the hands of the Ukrainian army, could have been used as they were the only weapons with the distinctive fragments. They insisted that the Russian army had phased them out, so it could not have been them – ignoring the likelihood that they would not have delivered their best equipment to the separatists, who were fighting to in East Ukraine for geopolitical independence so that they could join Russia.

Criminal proceedings subsequently were heard the District Court of The Hague in November 2022, in which European intelligence agencies had identified the suspects, and the Court found former Russian intelligence officers Igor Vsevolodovich Girkin and Sergey Nikolayevich Dubinskiy Guilty, along with Ukrainian separatist Leonid Volodymyrovych Kharchenko guilty of involvement in shooting the aircraft down. Another suspect, Oleg Yuldashevich Pulatov, was acquitted. The guilty Defendants were sentenced to life imprisonment and held jointly and severally liable for compensation of 16 million euros to the victims' next of kin. Of course, the Russian authorities had no intention of co-operating in the proceedings, and the Defendants were tried *in absentia*. Russia refused to acknowledge the authority or sovereign power of the Court over the Defendants and, in any event, continued its strategy to keep the world guessing about its involvement in East Ukraine. It did claim the defence of combatant immunity for the individuals, but this was rejected, and the Russian case was dismissed as lies. That being said, as the Russians could not possibly accede to a Guilty verdict, no appeal was lodged, and the sentencing was duly made final.

The Aftermath

This case illustrates the complexities involved in establishing the accountability of State actors, in which the facts and findings found in the technical investigation conducted in accordance with the Chicago Convention can simply be rejected without any compelling counter-evidence. It must, naturally, weaken the process of accountability, but the ultimate purpose of Just Culture remains intact: to make findings which will lead to an improvement in air safety, in which the entire range of professional practice will be involved.

The obvious option of re-routing can have an immediate effect in bringing the risk to acceptable levels, although there are some trouble-spots where the airlines are prepared to take a higher risk at management level; the actuaries will apply a cost-benefit model, and the option then remains open to transfer the risk to insurance.

Following Russia's invasion of Ukraine in 2022 and subsequent fears of escalating regional conflicts, Western States considered their options sanctioning Russian operators; the UK banned Russian airlines and any flights associated with a Russian person from operating in UK airspace or to UK airports, with the exception of transatlantic overflights. They were followed by European Union States, Canada and others; Russia responded with a predictable decision, banning flights from Russian airspace by the countries that have banned Russian flights.

Such action by the Western States may be described as preventative and punitive; but the wider question of the professional duty of care in air operations clearly must be raised. And a major factor arising out of the MH17 case involved the responsibility and accountability of Air Traffic Control. It is simple to justify the Just Culture model when applying the facts found in the MH17 case to the establish whether an individual fell below the standard demanded by the normative ethics of society; once again we are confronted with the criminalisation of air accidents by the creation of the offence of gross negligence. In specific terms, it was apparent that routing and Air Traffic Control had to meet very practical objectives to maintain the duty of care expected of them, to ensure, as reasonably as possible, the safety of civil aircraft when overflying hazardous areas. In other words, the process of continuous learning has an important effect on accountability.

THE MH17 CASE AND CONTINUOUS LEARNING

A key objective of Eurocontrol's Model Policy[10] is to provide some direction for the criminal investigation and prosecution of potential criminal offences resulting from aviation incidents or accidents that come to the attention of prosecutors so that both the safety investigation and judicial process can progress in parallel without either party acting in a manner prejudicial to the interests of the other. It was, therefore, entirely right and proper that both safety and judicial processes should be served in the light of the MH17 incident.

In the wake of the MH17 crash, an important conference took place in Amsterdam when Eurocontrol and the Dutch government hosted an international series of workshops on Just Culture, and the problems associated with gross negligence must be resolved. It was, therefore, fundamentally a forward-looking exercise, to identify problems and solutions that allow the two investigation processes to succeed, and minimise future risk.

This book has demonstrated something that confronted the conference, that the dynamics and deliverables in Just Culture are difficult to define. It does not matter whether the State follows the Roman system or English system of judicial procedure, the critical factor is to protect the investigation process. The State will rightly make the administration of justice the cornerstone upon which the State and its legal system are executed. We have seen how States have the rights and powers of sovereign jurisdiction, in which no other State can interfere; thus, it is important to ensure that the administration of justice meets the State's international imperatives, while balancing the normative ethics of society which often demands criminal accountability.

European Regulation 376 / 2014 sets out the general principles of Just Culture that will have direct effect across borders:

- A high general level of safety should be ensured in civil aviation in the European Union and every effort should be made to reduce the number of accidents and incidents with a view to ensuring public confidence in aviation transport.
- It reiterates the aim to prevent accidents by facilitating the prompt holding of efficient and high-quality safety investigations, but the Just Culture policy should not interfere with the process of accident and incident investigations managed by national safety investigation authorities as defined in the 2010 Regulation.[11]
- The sole objective of safety investigations should be the prevention of future accidents and incidents without apportioning blame or liability.

- According to the international standards and recommended practices set out in Annex 13 to the Chicago Convention, the investigation of accidents and serious incidents is to be conducted under the responsibility of the State where the accident or serious incident occurs, or the State of Registry when the location of the accident or serious incident cannot definitely be established as being in the territory of any State. A State may delegate the task of conducting the investigation to another State or request its assistance. Safety investigations in the EU should be conducted in a similar way.
- Article 15 addresses confidentiality and appropriate use of information:

> Member States, the Agency and organisations shall not make available or use the information on occurrences: (a) in order to attribute blame or liability; or (b) for any purpose other than the maintenance or improvement of aviation safety.
>
> Member States shall ensure that their competent authorities…. for the administration of justice cooperate with each other through advance administrative arrangements. These advance administrative arrangements shall seek to ensure the correct balance between the need for proper administration of justice, on the one hand, and the necessary continued availability of safety information.

No international regulation, of course, can interfere in the State's administration of justice; as a result, the Regulation highlights the need to strike the balance between the need for administration of justice, and the availability of safety information. The consequence raises issues of interpretation at the very start of a judicial investigation, namely the definition of gross negligence at the State level, and how that maps onto the definition of other states. Just Culture promotes continuous learning, certainly, for that imperative crosses borders, but achieving the balance with the administration of justice may prove problematic on two grounds:

- The definitions of criminal accountability vary between States and therefore defy a common application.
- A State may seek to justify decisions in the administration of aviation within its airspace in order to meet political imperatives, which have no place in a Just Culture and distort fairness in accountability. A stark example took place in 2021.

Case Study: Ryanair Flight 4978

In the light of these cases, let us revisit the incident which we discussed in Chapter 3, and consider the consequences for State accountability.

It will be recalled that this was a regularly scheduled international passenger flight from Athens to Vilnius operated by Buzz, a Polish subsidiary of the Irish airline Ryanair.

The incident in question concerned an aircraft carrying dissident Belarusian journalist Roman Protasevich on 23 May 2021. While in Belarusian airspace, the pilot was ordered to divert from its course and to land in Minsk by Belarusian Air Traffic Control on instructions from the government alleging that a bomb threat on the aircraft had been received. The flight crew had no prior suspicion of this, but in accordance with their duty of care for air safety, they followed instructions diverting the aircraft, to land at Minsk; in case the pilot in command had any doubts about the action he should take, a Belarusian fighter jet approached and shepherded it into Minsk.

After the aircraft had landed, Belarusian authorities searched the baggage on board, and checked the identities of passengers. The claim lacked any corroborative evidence of a bomb on board; and if there had been a presumed risk to the safety of the aircraft, given that it was closer to Vilnius than to Minsk, it made no sense to divert it. The conclusion was that this must have been ruse; there had not been any bomb threat, which the Belarusian government authorities had falsely invented as a means to take control over the flight while in its airspace, in order to seize two dissidents. They duly arrested the two passengers and took them into custody: Protasevich, a prominent opponent of Belarusian dictator Alexander Lukashenko, and his girlfriend, law student Sofia Sapega.

Obviously Belarus adopted the ruse in order to force the pilot in command to land, for the safety of air navigation, and the ruse betrayed their unlawful motive, defying international law. Belarus acceded to the Chicago Convention in 1993, which contains important provisions in this case.

Article 1 states *Every state has complete and exclusive sovereignty over airspace above its territory.*

Article 3bis states *The contracting States recognize that every State, in the exercise of its sovereignty, is entitled to require the landing at some designated airport of a civil aircraft flying above its territory without authority or if there are reasonable grounds to conclude that it is being used for any purpose inconsistent with the aims of this Convention…*

Article 5 states *The aircraft of states, other than scheduled international air services, have the right to make flights across state's territories and to make stops without obtaining prior permission. However, the state may require the aircraft to make a landing.*

Article 16 states *The authorities of each state shall have the right to search the aircraft of other states on landing or departure, without unreasonable delay.*

Protasevich and Sapega were still in the aircraft when detained by the Belarusian authorities, which was registered in Poland and thus, they were protected by the European Convention on Human Rights – in theory. In any event, no country has the right to detain suspects on a civil aircraft for crimes that were not committed on board that aircraft. As a result, even though the aircraft and crew were not detained, Poland was obliged to seek the release of the detained passengers. If Belarus felt justified in seeking custody of them, it should have should have applied to Poland or to Lithuania (as Vilnius was the port of destination) for their extradition for alleged crimes in Belarus. Consequently, Poland could also bring a claim against Belarus under this head.

The incident naturally created a great deal of alarm and international outrage, ultimately calling for proceedings to be instigated against Belarus in the International Court of Justice. It is a fact of life, however, that such proceedings are dreadfully slow and, ultimately, lack meaningful punishment. There is a possibility that such a consequence might have been contemplated by Belarus but, clearly, it proved a highly acceptable risk, in taking its actions in order to meet political imperatives, which have no place in a Just Culture and distort fairness in accountability.

However, collateral action was taken by the US Department of Justice,[12] which claimed some jurisdictional interest as four American citizens had been on the flight. Investigations followed by the FBI's New York Field Office, Counterintelligence Division Foreign Influence Task Force and the New York Joint Terrorism Task Force, which principally consisted of special agents from the FBI and detectives from the NYPD; as well as the FBI Legal Attaché Offices in Riga, Latvia; Warsaw, Poland; Athens, Greece; Kiev, Ukraine; and London; the National Transportation Safety Board; the Justice Department's National Security Division Counterterrorism Section and Criminal Division's Human Rights and Special Prosecutions Section; and the Office of International Affairs. Poland's Internal Security Agency, Mazowieckie Regional Prosecutor's Office and Lithuanian authorities provided valuable assistance with this investigation.

Following the investigation, a federal grand Jury in New York returned an indictment charging four Belarusian state officials with conspiracy to commit aircraft piracy. The individuals consisted of the Director General of the Belarusian state air navigation authority, and his deputy, who allegedly conspired to communicate the false bomb threat to staff at Minsk Air Traffic Control, directing them to instruct the Flight divert to Minsk in response to the purported threat. The two others were officers of the Belarusian state security services, who were allegedly instrumental in the execution of the plot.

It would come as no surprise that the Department of Justice concluded its statement thus:

> The defendants remain at large. The United States looks forward to working with our foreign partners to bring them to justice.

In light of events in Ukraine less than a year later, the process remains unresolved, and Belarus has a great deal more to worry about than claims of criminal accountability in the case. In the meantime, civil aircraft have vanished from routes overlying Belarus, and Belarusian aircraft have been banned from European States.

These cases demonstrate the importance of understanding how fairness actually works in the reality of dealing with state actors. The continuing lessons must be learned and applied by aviation professionals, whether they are pilots, flight routers, Air Traffic Controllers, or senior managers, in the steps which they must take in the risk management process arising out of such scenarios.

State authorities inevitably must consider the risk management function in their operations, but they may not be willing or able to control the mischief for which they are accountable. Accountability necessarily follows a retrospective pattern – we learn from our mistakes – but, while this continuous learning process constantly informs and evolves the civil operators' safety risk management, it is much more difficult to abate the risk that we have seen in the cases of Iran Air Flight 655 or Malaysian Flight 17, when civil aircraft are shot down by State actors.

As a post script, it must be said that even more recent experience underlines the consequences of such State mischief. On 8 January 2020, Ukraine International Airlines Flight 752 was operating blamelessly on a scheduled flight from Tehran to Kyiv, when the Boeing 737-800 was shot down by the Iranian Islamic Revolutionary Guard Corps (IRGC) shortly after take-off from Tehran Airport, killing all 176 people on board. After initially denying responsibility, two days later, Iran was forced to accept the findings of international investigations – and public protest vis social media – when it admitted that the IRGC had unintentionally targeted the aircraft which they mistook for a hostile cruise missile. In mitigation, they said that the IRGC had shot down the Ukrainian jet by accident, the result of human error, while the force was on high alert, and mistook it for a hostile missile and made a bad decision. Given the crimes committed by the IRGC in maritime operations against Western shipping, such confessions sound somewhat hollow.

Of course, no matter what is offered in mitigation, the civil aviation industry has to respond to the threats contained in incident reports. Given the problem in which a missile will hit a civilian aircraft virtually without warning, research and development must design a solution, upon which pilots can rely to inform their risk assessment. Saab Avitronics, Chemring Countermeasures and Naturelink Aviation[13] co-operated to develop the Civil Aircraft Missile Protection System, an infrared countermeasure against infrared-homed anti-aircraft missiles, specifically designed to defend civilian aircraft flying under 15,000 feet against portable surface-to-air missiles. Emergent technologies in such civil aircraft defence are slow, though, possibly because the cost-benefit decisions of airlines to buy such protection may result in decisions to adopt the cheaper alternative of avoiding the trouble-spots in routing. Responding with solutions in the continuous learning process can have its limits when it comes to the economics of air operations.

Given the experiences in these case studies, equipping civil airliners with the means of self-defence appears to be such an obvious option that the surprise is that it has not been adopted before – until one factors in the point that such risks may be regarded by airlines as remote contingencies and do not merit the investment on a cost-benefit analysis. The increasing incidents of military attacks on civilian aircraft may necessitate a re-examination of such decisions, though, if the risk strays more closely towards recklessness. Moreover, the question inevitably arises: What would the insurers say?

THE IMPORTANCE – AND DANGERS – OF THE PILOT'S DISCRETION

It is worth repeating, that domestic legislation regulates the responsibility of the Pilot in Command. The headline statute is the Civil Aviation Act 1982, which implements the provisions of the Chicago Convention 1944,[14] whose preamble sets out its purpose *to lay down principles and make arrangements in order that international civil aviation may be developed in a safe and orderly manner and that international air transport services may be established on the basis of equality of opportunity, and may be operated soundly and economically.*

The Convention is supported by 19 annexes containing standards and recommended practices (SARPs). The Convention runs parallel with UNCLOS in that it provides under Article 1 (Territorial Sovereignty) that *the contracting States recognize that every State has a complete and exclusive sovereignty over the airspace above its territory.*

Article 17 addresses the Nationality of aircraft and provides that *aircraft have the nationality of the State in which they are registered.* Thus, aircraft have prima facie the protection of the registered State as if it were sovereign territory and the State is also reciprocally responsible for the international good conduct of its aircraft as regards other States; this, after all, was the bedrock of Poland's position in the Ryanair case. Airline nationality is not actually addressed in the Convention, although it has become an important part of bilateral air agreements, as well as the multilateral Transit agreements, which do address ownership in terms of the legal body which exerts substantial ownership and effective control over the aircraft, with the parallel of Offshore Flags in maritime operations.

Article 44 of the Chicago Convention stipulates that the overall objective of the International Civil Aviation Organisation is to ensure the safe and orderly growth of international civil aviation throughout the world, as such managing the regime envisioned by the Convention. The latest revision to Annex 13[15] emphasises the need to develop appropriate legal guidance that will assist States to enact national rules to protect information from safety data collection and processing systems (SDCPS), while allowing for the proper administration of justice in the State; but it cannot dictate to the sovereign regimes of those States how they must define and enforce their own criminal legislative processes, as a result of which international regulation can determine the parameters of a Just Culture to its heart's content, but cannot influence the administration of justice for the prosecution of offences.

Once again, it is worthwhile comparing the position of the Pilot in Command with the ship's Master. The United Nations Convention on the Law of the Sea[16] (UNCLOS) has something to say in both scenarios, Article 2 of which assures the State its sovereign jurisdiction, to the exclusion of other States, *from its land territory to the territorial sea up... and to the air space over the territorial sea as well as to its bed and subsoil.* The Master retains the ancient right of absolute discretion, in a rule far predating any international convention but enshrined in SOLAS, that nobody shall *prevent or restrict the master of the ship from taking or executing any decision which, in the master's professional judgement, is necessary for safety of life at sea and protection of the marine environment.*[17]

While the shipmaster's absolute discretion is very wide, it is notable that this is not so closely paralleled with the authority of the pilot in command, which is more narrowly drawn. The Civil Aviation 1982 gives the authority for the Air Navigation Order 2016,[18] which provides under Article 68:

> The pilot in command of an aircraft is responsible –
>
> a. before every flight, for defining the roles and duties of each crew member;
> b. for the operation and safety of the aircraft and for the initiation, continuation, termination or diversion of a flight in the interest of safety; and
> c. during aircraft operations, for ensuring the safety of all crew members, passengers and cargo on board.

The discretion of the pilot, therefore, is more sharply focused, and Article 69 reduces the discretion in departing from established procedures:

> The pilot in command must ensure that –
>
> a. the flight is performed in such a way that the operating procedures specified in the flight manual, or where required the operations manual, for the preparation and execution of the flight are followed; and
> b. procedures are established and followed for any reasonably foreseeable emergency situation.

Case Study: Robert Murgatroyd

This case illustrates how the pilot's responsibility for air safety goes far beyond the command and control of the aircraft, embracing the wide scope of the Air Navigation Order 2016 (as amended). In particular, Chapter 2 (68) defines the responsibilities and authority of the pilot in command, inter alia providing that the pilot is responsible for *the operation and safety of the aircraft and for the initiation, continuation, termination or diversion of a flight in the interest of safety; and during aircraft operations, for ensuring the safety of all crew members, passengers and cargo on board.* In addressing operating procedures, subsection 3 of the provision following adds that *the pilot in command must ensure that the flight is performed in such a way that the operating procedures specified in the flight manual, or where required the operations manual, for the preparation and execution of the flight are followed.* Subsection 9 provides for mass and balance requirements, stating *The pilot in command must ensure that during any phase of operation, the loading, the mass and... the centre of gravity position of the aircraft comply with any limitation specified in the flight manual, the weight schedule.... or equivalent document.*

Moreover, Article 240 defines the very serious offence of endangering the safety of an aircraft: *A person must not recklessly or negligently act in a manner likely to endanger an aircraft, or any person in an aircraft.*

Fifty-two-year-old Robert Murgatroyd was a very experienced pilot having held a private pilot's licence for many years, but held neither a Commercial Pilot's Licence nor an Air Operator's Certificate, contrary to the Air Navigation Order for fare-paying passengers. On 9 September, 2017, he took off in a 45 year old Piper PA-28 from Manchester Barton Aerodrome. It was a non-scheduled flight, but nevertheless was a public transport flight as the passengers paid him a fee for the leisure trip from Barton to the Isle of Barra in the Outer Hebrides. Without a commercial pilot's licence nor an Air Operator's Certificate, of course, he did not have any valid insurance. The three passengers paid £500 each for the trip, and he filled the fuel tanks to capacity; he made no attempt to weigh the passengers or their luggage before making the required checks on the weight of the aircraft and with the wrong flight manual on board. In fact, the aircraft was grossly overloaded, at 426 pounds over the maximum take-off weight of 2,150 pounds. He then failed to give any safety briefing.

The weather report was wet and 'very poor,' but the Aerodrome manager had made an inspection to check the grass runway at Barton and decided that it was still usable.

Murgatroyd performed a practice take-off run, to check how well the aircraft could accelerate on the runway. On eye-witnessed, an experienced flying instructor gave evidence that the front wheel of the aircraft showed no sign of lifting off the ground. The pilot then turned around to perform another practice take off, and the eye-witness described the acceleration as sluggish. This time however, the aircraft took off. Ground witnesses were concerned, however, and the aerodrome's fire crew put on their protective firefighting kit in readiness. Witnesses said that the aircraft was struggling in the air; then, as it was climbing, it was forced to turn left to avoid hitting electricity pylons, obstacles which would have proved extremely hazardous to a struggling overloaded aircraft. He then reassessed his flight path and doubled back over the M62, hazarding people and property on the ground again. It then clipped the trees at the side of the M62, before crashing into a field about 160 yards from the motorway. The left wing was ripped off, causing fuel to spill out, and the undercarriage was also ripped off.

Mercifully, there were no fatalities. The pilot suffered a broken nose, one passenger suffered a cut to his hand; the second suffered suspected cracked ribs and whiplash. The third passenger suffered a serious cut to the head and severe bruising to his ribs; he stated, *I banged my head. I didn't know where the blood was coming from at first. I was panicking. I was covered in blood. I could then smell petrol and then obviously I am trying to get out of the door.*

In a victim impact statement read to the court, he said, *I knew we were going to crash. I am not a dramatic person but generally people do not survive plane crashes. I distinctly recall thinking 'we have had it.' It was absolutely petrifying. The pilot did nothing to reassure us or prepare us for what was happening. He didn't even tell us to brace. Then we crashed. It's a bit of a blur but somehow I managed to get the seat belt off and out of the door. I have got blood all over me. I don't know how the others got out. I didn't see them at all. One of the other lads, bless him, came to help.*

Hardly more compelling evidence was needed for the safety briefing.

Murgatroyd pleaded Not Guilty to a litany of charges:

- Recklessly endangering the safety of an aircraft or persons in an aircraft
- Recklessly endangering the safety of persons or property
- Conducting a public transport flight without an Air Operator Certificate
- Acting as a pilot without holding an appropriate licence
- Flying outside the flight manual limitations
- Flying without insurance

In his interview under caution, he denied recklessness, saying that, at worst, it was a genuine mistake; in fact he went further, describing himself as a hero for the way he handled the forced landing and said that there should be a Hollywood film made about him.

As for the commercial nature of the flight, Murgatroyd pointed out the provisions of cost-sharing in European and National regulations, which permit cost sharing flights in the following circumstances, namely that the direct costs of the flight must be shared between all of the occupants of the aircraft, including the pilot, up to a maximum of six persons, who must be private individuals.[19] Direct costs are defined as the costs directly incurred in relation to a flight, including fuel, airfield charges, rental fee for an aircraft, but there can be no element of profit. His problem was that the fees paid by the passengers exceeded the profit threshold by £1,000 and, in any event, there was no evidence that he had joined in a fee-sharing agreement.

The Jury considered all the evidence and found him Guilty on all seven counts. Sentencing him, the Judge said, *You are a cavalier operator, and effectively an air accident waiting to happen.*[20] He was sentenced to a term of imprisonment of forty two months. We have discussed a number of cases in which the fairness of the case was in doubt, compromising the policy of Just Culture. In this case, the Prosecutor was very confident that the Defendant's conduct had crossed the line between blameless negligence and criminal accountability; and the Jury had no doubt either.

Offenders should not necessarily breathe a sigh of relief if their cases are heard in the lower Court, the Magistrates Court. While there had been an upper sentencing limit of £5,000 for Air Navigation Order offences, since 2015 for such Level 5 offences, Magistrates have the power to impose fines of an unlimited amount. The stated purpose for the change was to allow the financial means of the defendant to be reflected better in the punishment. Of course, it also permitted an enormous cost-saving measure on the Court process, enabling cases to be concluded in the Magistrates Court without the expense in time and money of committing the case to the Crown Court for sentence.

While the aim of Just Culture is to promote continuous learning from previous mistakes and to encourage pilots to openly and freely share essential safety related information, it does not restrain the criminal process from prosecuting individuals whose violations of air safety laws are manifestly wilful or reckless, endangering the safety of the aircraft or persons on the ground.

Article 240 regulates, rather, by prohibition, conduct endangering the safety of the aircraft, in much the same way that Section 58 Merchant Shipping Act 1995 prohibits conduct endangering the safety of life at sea, applying to aircraft registered in the UK, apply to such aircraft wherever they may be; and such other aircraft when they are within the UK or on or in the neighbourhood of an offshore installation:

> (2) Except where the context otherwise requires, the provisions of this Order in so far as they prohibit, require or regulate (whether by express reference or otherwise) the doing of anything:
>
> (a) by persons in, or by any of the crew of, any aircraft registered in the United Kingdom, apply to such persons and crew, wherever they may be;

A European-wide plan for Standardised European Rules of the Air (SERA) have been brought into line with UK law following the Rules of the Air Regulations 2015, which were made in consequence of the coming into force in the UK of Regulation (EU) No. 923/2012 of the European Parliament and of the Council of 26 September 2012 laying down the common rules of the air and operational provisions regarding services and procedures in air navigation.

SERA.2010 Responsibilities

a. Responsibility of the pilot-in-command

   The pilot-in-command of an aircraft shall, whether manipulating the controls or not, be responsible for the operation of the aircraft in accordance with this Regulation, except that the pilot-in-command may depart from these rules in circumstances that render such departure absolutely necessary in the interests of safety.

b. Pre-flight action

   Before beginning a flight, the pilot-in-command of an aircraft shall become familiar with all available information appropriate to the intended operation. Pre-flight action for flights away from the vicinity of an aerodrome, and for all IFR flights, shall include a careful study of available current weather reports and forecasts, taking into consideration fuel requirements and an alternative course of action if the flight cannot be completed as planned.

Perhaps most critical to the authority of the Pilot in Command is the provision in the Official Journal of the European Union (13.10.2012) regarding *Authority of pilot-in-command of an aircraft:*

> The pilot-in-command of an aircraft shall have final authority as to the disposition of the aircraft while in command.

The UN has adopted standards through the ICAO; Annex 6, Part II provides:

> 3.2 The pilot-in-command shall be responsible for the safety of all crew members, passengers and cargo on board when doors are closed. The pilot-in-command shall be responsible for the operation and safety of the aeroplane from the moment the aeroplane is ready to move for the purpose of taking off until the moment it finally comes to rest at the end of the flight and the engine(s) used as primary propulsion units are shut down.

As a result, all the signatories must implement these provisions within their domestic law.

US legislation provides under CFR 14 § 91.3 Responsibility and authority of the pilot in command:

a. The pilot in command of an aircraft is directly responsible for, and is the final authority as to, the operation of that aircraft.
b. In an in-flight emergency requiring immediate action, the pilot in command may deviate from any rule of this part to the extent required to meet that emergency.
c. Each pilot in command who deviates from a rule under paragraph (b) of this section shall, upon the request of the Administrator, send a written report of that deviation to the Administrator

If there is a parallel with the shipmaster's absolute discretion, it is here.

Such statutory provisions establish the framework of responsibility, and international legislation in recent years has sought to rationalise this framework in the context of the priority needed for flight safety over a priority to criminalise offenders in State courts – but the reality of criminal law, evidence and practice has drifted far from this, with the result that a gulf has emerged between the theory of justice envisaged by international Conventions and the reality of justice which is demanded by the normative ethics of Society, in other words, the process which is served by the State's sovereign judicial function. Mildred Trögeler[21] rightly identified this as one of the main obstacles in resolving a solution to the problem of criminalisation, making the point that States are reluctant to change their national legislation, although her observation that individual States have not exercised the option to delegate their sovereignty function is hard to follow, as most States do not have the option to erode their sovereignty, which would be constitutionally unacceptable. It would also be difficult to reconcile with Article 2 of the UN Charter, which embraces the principle of the sovereign equality of all its Members. Her solution to change the law at the national level may well abate the mischief, in so far as concepts of criminal negligence and associated strict liability would be abandoned, but such a solution may be likened to the observation of a Judge who was presiding over a case in which a colleague of the author's asked a witness under cross-examination whether a particular contingency were possible. The Judge said:

> That is not the proper question. It is possible that, at any moment, a polar bear could come through that door. The proper question should be: is it likely?[22]

The core feature of the problem thus remains the fact that, whatever international regulation is attempted, the judicial regime remains jealously guarded as the cornerstone of State sovereignty, which has perpetuated the criminalisation phenomenon through the emergence of criminal negligence. It is a feature shared between shipmasters, pilots in command, air traffic controllers and beyond; nowhere more relevant can this be found, than in the exercise of their professional discretion. As we have seen, the Master has the comfort of knowing that no other person can prevent or restrict them taking any decision which, in their professional judgment, is necessary for the safety of life at sea; while the Pilot in Command is responsible for the operation of the aircraft in accordance with Regulations, but may depart from them if absolutely necessary in the interests of safety. Of course, they had better have good evidence for their decision, because, if they get it wrong, they will be held accountable.

Case Study: Olympic Airways Flight 411

Olympic Airways operated a flight from Athens Ellinikon International Airport bound for John F Kennedy Airport in the United States. On the day in question, the pilot in command of the Boeing 747 was 55-year-old Sifis Migadis, who had 32 years of experience with Olympic. The co-pilot, Konstantinos Fikardos, was equally experienced and was a close friend of Migadis, as well as being flight instructor.

The departure airport was surrounded by a complex geography. Athens lies in the centre of four large mountains, to the north, northeast, east and west, which is responsible for cause a temperature inversion phenomenon, which develops when cool air is trapped at the ground under a layer of warm air, and pilots must take this into account. On the day in question, the ground temperature was 32 degrees centigrade, which presented problems for aircraft engines which were idling awaiting take-off.

The aircraft in question, named *Olympic Zeus*, was a model which featured more powerful engines and a higher maximum take-off weight the previous 747-100 model. The high-bypass turbofan had been developed in the late 1960s by Pratt & Whitney; it featured a water-injection system that delivered more thrust, particularly valuable in taking off. Pratt & Whitney had been under pressure at the research and development stage, however, to meet customer demand, which resulted in problems with the engine that had not been rectified, particularly with a number of mechanical issues and proper maintenance was required to ensure the engine's safety. Indeed, the engine had already suffered incidents, including a serious casualty in 1971, when the 747's number thee engine exploded after its turbine blades disintegrated. The National Transportation Safety Board stated in the casualty report that this was likely due to an ongoing pattern of the engine overheating during starting procedures; given the climate conditions at Athens in our case, this would carry considerable weight in the case.

On 9 August 1978, the flight with 418 passengers and crew was scheduled to take off from Athens at 14.00 on a direct flight to New York. The plane was heavy with the 160 tons of fuel required for the transatlantic flight. The aircraft thus weighed 770,000 pounds on the day of the flight, which compared badly with the maximum weight permitted for take-off climb at this airport limited to 353,000–778,000 pounds for the wet thrust engine. The aircraft had been cleared for take-off and had reached the critical speed for rotation when an engine exploded; having passed the critical moment, however, the flight crew had no alternative but to take off. It was a very public scene, as the airport Air Traffic Control actually watched it trying to gain height, with engine debris littering the ground. Then the cockpit waterflow indicator flashed, and in the crisis moment the flight crew mistakenly thought it was emptying, so they switched off the water pump. But with just three engines, they were struggling to climb at low altitude; then engine number two had a technical failure; it had been serviced before the flight, but was not fully operational.

The pilot in command had about 90 seconds to manage the risk. He duly determined the risk, decided to treat it with an action plan, then within seconds prepared the action plan using his own initiative and discretion, and subsequently monitored and reviewed the action plan.

Clearly, he had to assess the airspeed and altitude risk; then, to try to mitigate the danger, he decided to retract the undercarriage, in order to reduce drag. At the very low height they had reached, this was a violation of the regulations according to the aircraft's instructions; but the captain relied, instead, on his experience and discretion in making the decision. To danger was that, to retract the undercarriage, the doors would increase drag, but he considered that this was the lesser of two evils, in order to avoid hitting the mountain at that height. He eventually managed to bring the nose up and cleared the mountain; but they did not have enough speed, and he only just managed to avoid the aircraft stalling.

Another hazard was confronting them, for they were flying low over an area of 100,000 inhabitants; crashing now would be catastrophic to them. The crew had to fly 200 feet at a speed close to stalling, when the aircraft would lose lift and control. The flight engineer then managed to get the other three engines working optimally and with a coastal breeze now assisting gave them lift. Captain Migadis had just enough to bank and turn towards the sea, where he climbed to 1,500 feet, and he could jettison fuel to lighten the load and reduce fire risk.

Incredibly, the flight returned to Athens with no casualties. With rapturous applause on the ground, and enormous praise for the flight crew's flying skills.

The Air Accident Investigation and Aviation Safety Board, the Greek national air safety board that investigates aviation accidents and incidents, had not been established at that time, leaving Boeing the room to defend itself, in a report that may be open to verification. Brien Wygle, the vice president of Customer Support at Boeing, issued a report[23] which devalued Captain Migadis's effective management of the risk, stating that *the aircraft was recovered and climbed out normally by nosing over and accelerated with throttles advanced to approximately go-around thrust....The landing gear events are purely speculative.* The report concluded:

> No malfunction in airframe or engines caused the lack of performance following take-off. Instead, the problem was caused by the inadvertent shut off of the water injection pumps by the flight crew and the resulting decrease in thrust. Loss of wet thrust in a situation where the airplane was take-off climb limited severely reduced the ability to continue the flight with any significant amount of positive climb gradient. Once thrust was manually increased at a coordination time of approximately 325 seconds, the airplane climbed out in a normal manner.

Unfortunately for Boeing, experts were able to rely on evidence which conflicted with their conclusion denying that the cause was primarily due to engine failure. Newspapers reported that engine three exploded during take-off due to overheating of the turbine cooling pipes, a problem which had been encountered previously with such engines, and engine number two was performing badly before the flight engineer had managed to restore performance. In fact, all Boeing simulations of the flight resulted in crashes, so one is left asking the question, how did the aircraft survive? Nevertheless, Olympic Airways changed some of its procedures based upon the lessons learned from this flight, thus demonstrating the value of continuous learning.

CORPORATE ACCOUNTABILITY

In Chapter 1, we discussed the basic principles of corporate accountability and, indeed, Boeing's exposure to the risks of corporate accountability in recent cases, notably the Alaska Airlines crash in January 2024. In the following March, Dave Calhoun, Boeing's Chief Executive Officer, announced that he would step down at the end of the year, adding in his statement that the Alaska Airlines saga was *a watershed moment for Boeing.* The FAA's audit had found a litany of issues in Boeing's production processes resulting in an order to rectify the problems within 90 days. Calhoun concluded in a very fair mitigation:

> We must continue to respond to this accident with humility and complete transparency. We also must inculcate a total commitment to safety and quality at every level of our company. The eyes of the world are on us, and I know we will come through this moment a better company, building on all the learnings we accumulated as we worked together to rebuild Boeing over the last number of years.

The problems besetting the Boeing 737 Max were redoubled in the Lion Air and Ethiopian crashes, leading American attorneys alleging in June 2024 that the Department of Justice was putting pressure on Boeing to plead Guilty over its involvement in those two crashes in which 346 people died.

It is perhaps appropriate that we expand somewhat on the discussion of the modern English Statute Law of corporate accountability that we started in Chapter 1. On 6 April, 2008, long-awaited legislation came into force, which would change the face of corporate accountability for manslaughter. From sullen grumblings as far back as 1912 in the wake of the *Titanic* disaster, there had been growing concern that corporate manslaughter directly or indirectly threatens the security or well-being of society, and that it is not safe to leave it to be redressed only by compensation in civil proceedings. Following the failure of the prosecution of the shipowner in the case of the *Herald of Free Enterprise*,[24] the call for a change in the law demanded action. In the prosecution of Great Western Trains for corporate manslaughter following the Southall crash, the Tial Judge, Scott Baker J, ruled that the Defendant Company could only be convicted via the guilt of a human being. The Trial Judge identified that causal link with the controlling mind of the Company in the person of Richard George, the managing director responsible for safety matters, but causal link existed.[25] Jonathan Caplan, QC, instructed by Great Western Trains, was aware of the weakness and informed the Court that they would defend the charge of Corporate Manslaughter in the Indictment but would plead Guilty to a strict liability offence under the Health and Safety at Work Act, alleging that the Defendant Company exposed passengers to risks to their health and safety, which would end up with no more than a fine.

The Law Commission presented a report with recommendations, parts of which the government embraced,[26] which culminated in the Corporate Manslaughter and Corporate Homicide Act 2007.

The Offence of corporate manslaughter is created by Section 1, which states in the following subsections that an organisation, whether a corporation or even a government department, is guilty of an offence if the way in which its activities are managed or organised causes a person's death, and amounts to a gross breach of a relevant duty of care owed by the organisation to the deceased.

Subsection 3 provides that an organisation is guilty of an offence under this section only if the way in which its activities are managed or organised by its senior management is a substantial element in the breach referred to in subsection (1), while subsection 4 defines some key provisions:

For the purposes of this Act –

a. 'relevant duty of care' has the meaning given by section 2, namely:
    1. A 'relevant duty of care,' in relation to an organisation, means any of the following duties owed by it under the law of negligence –
        a. a duty owed to its employees or to other persons working for the organisation or performing services for it;
        b. a duty owed as occupier of premises;
        c. a duty owed in connection with –
            i. the supply by the organisation of goods or services (whether for consideration or not),
            ii. the carrying on by the organisation of any construction or maintenance operations,
            iii. the carrying on by the organisation of any other activity on a commercial basis, or
            iv. the use or keeping by the organisation of any plant, vehicle or other thing;
        d. a duty owed to a person who, by reason of being a person within subsection (2), is someone for whose safety the organisation is responsible.

Subsection 4 continues

b. a breach of a duty of care by an organisation is a 'gross' breach if the conduct alleged to amount to a breach of that duty falls far below what can reasonably be expected of the organisation in the circumstances;

c. 'senior management,' in relation to an organisation, means the persons who play significant roles in –
   i. the making of decisions about how the whole or a substantial part of its activities are to be managed or organised, or
   ii. the actual managing or organising of the whole or a substantial part of those activities.

The act provides that the offence triable by indictment only, before a Jury which is given as much guidance as possible in the factors to consider

1. This section applies where –
   a. it is established that an organisation owed a relevant duty of care to a person, and
   b. it falls to the Jury to decide whether there was a gross breach of that duty.
2. The Jury must consider whether the evidence shows that the organisation failed to comply with any health and safety legislation that relates to the alleged breach, and if so –
   a. how serious that failure was;
   b. how much of a risk of death it posed.
3. The Jury may also –
   a. consider the extent to which the evidence shows that there were attitudes, policies, systems or accepted practices within the organisation that were likely to have encouraged any such failure as is mentioned in subsection (2), or to have produced tolerance of it;
   b. have regard to any health and safety guidance that relates to the alleged breach.
4. This section does not prevent the Jury from having regard to any other matters they consider relevant.
5. In this section 'health and safety guidance' means any code, guidance, manual or similar publication that is concerned with health and safety matters and is made or issued (under a statutory provision or otherwise) by an authority responsible for the enforcement of any health and safety legislation.

Notwithstanding the statutory definition, as a matter of practice, the Judge will often direct the Jury in accordance with the language used in respect of the common law offence of gross negligence manslaughter. Various formulas have been used, including *truly, exceptionally bad* and *so bad that it amounts to a crime and deserving of punishment*; here we are back with the case of *Adomako.*[27] Overall, the duty of the Judge is to bring home to the Jury the very high threshold which amounts to gross failure.

It is apparent that Section 8 seeks to mitigate the capricious nature of Jury deliberations over which nobody has any control, with guidance the following key issues:

- whether the defendant failed to comply with any health and safety legislation;
- if so, how serious that failure was;
- how much of a risk of death the failure posed;
- the extent to which the evidence shows that there were attitudes, policies,
- systems or accepted practices within the organisation that were likely to have
- encouraged any such failure; and
- the contents of any health and safety guidance that relates to the alleged breach.

For all such guidance, Section 8(4) also makes it clear that this does not preclude the Jury from taking into account any other factors which they may consider relevant. For example, evidence of a previous Near miss incident, a failure to follow the organisation's own written procedures, or evidence of cost cutting may all be relevant.

Parliament had to consider a vital factor in the Corporate Manslaughter Bill; namely, that the offence had to distance itself from the accountability of an individual for a stand-alone offence. Individuals are animate beings whose accountability depends on considerations personal to them:

- The Defendant must have owed a duty of care to the deceased; and
- The Defendant must have been in breach of that duty; and
- The breach was so grossly negligent that it should be seen as criminal.

The individual's duty of care requires foresight, in other words, the reasonable apprehension that there was a serious and obvious risk of death, from which the rest of the test must follow. However, a corporation is an inanimate body which does not have the capacity to foresee risk. Such an issue, therefore, is not within the Jury's contemplation in corporate accountability. The Jury's task is a complex one, however, because evidence that the risk was, or should have been, obvious to senior managers within the organisation would be evidence which the Jury could take into account in determining whether there had been a gross breach as defined in Section 1(4).

Case Study: CAV Aerospace

Prosecutions of aviation-related cases have been limited (so far) to the case of CAV Aerospace Limited.[28] This was the parent company of CAV Cambridge, based at Cambridge Airport, which employed the deceased, who was working on the site where a stockpile of metal billets had been stacked exceeding safe height limits, as a result of overstocking and insufficient space. On 26 January, 2013, the stack collapsed and crushed the victim.

The parent company was responsible for matters such as purchase orders and delivery, as a result of which CAV Cambridge had no say in the management process of the stock. CAV Cambridge had already identified the risk, and six near miss reports had previously been submitted to CAV Aerospace, none of which had resulted in any action or risk assessment being taken. The Court held that the parent company had to accept significant responsibility for the accident.

At the Trial, CAV Aerospace accepted that a serious Injury was reasonably foreseeable. It did not accept responsibility, however, arguing that the reason for its failure to heed warnings and act was that it did not know it was responsible for the safety of the employees of CAV Cambridge; in other words, it did not accept responsibility for the management failure. However, the corporate structure weaved quite a tangled web, as the sole director of CAV Cambridge was also a director of the parent company, who often ignored emails from the general manager of CAV Cambridge, and managers at Cambridge had raised the unsafe system of work in question with senior management at CAV Aerospace several times during the three years preceding the fatal incident. As if that were not enough, a formal Health and Safety warning had brought to the Company's attention the danger arising out of the stacked billets. It appeared, however, that solutions were rejected on cost grounds and CAV Aerospace did not appear to agree it had a duty of care to the employees of CAV Cambridge.

The Jury found the Defendant Company Guilty of Corporate Manslaughter. In sentencing, the Judge decided not to consider cost cutting an aggravating factor, but the factors that most aggravated the offence were the lengthy duration of the failures, the level of the failures and the failures to heed warnings. On the other hand, the judge considered that there were mitigating factors, namely that the Defendant co-operated with the investigation, subsequently taking remedial action, and paying compensation to the deceased's family above the level of an insurance payment.

The Court imposed a sentence of a fine of £600,000, as well as and £400,000 for Health and Safety strict liability offences that also appeared on the indictment. It was the largest company to be convicted of corporate manslaughter at that date.

The key question posed by the act regarding the risk assessment of senior management is whether the way in which activities were managed or organised by senior management contributed as a substantial element in the facts of the case; the burden is upon the Prosecution to prove the case beyond reasonable doubt. Clearly, even though CAV Aerospace was the parent company, the Court was satisfied that they owed the relevant duty of care in the case and, accordingly, the senior management failure was theirs, playing significant roles in the making of decisions about how the whole or a substantial part of its activities are to be managed or organised, or the actual managing or organising of the whole or a substantial part of those activities. Possibly without realising it, the Court underlined the importance of Just Culture in the process of continuous learning from previous mistakes, and the importance of improving safety and work practices in the light of warnings, even if, as in this case, the fatal accident was not strictly related to an occurrence associated with the operation of an aircraft. The downstream consequence on the offender will be just the same in a criminal prosecution, in which the Court must be satisfied beyond reasonable doubt that the key elements of the *mens rea* and *actus reus* have been met for corporate manslaughter, which does not differentiate between aviation and non-aviation cases. If the Company causes a person's death, and their conduct amounts to a gross breach of a relevant duty of care owed by the organisation to the deceased, then the act will apply, whether on the ground, or in the air, within the sovereign jurisdiction of English law.

Other jurisdictions will apply their own laws, of course. There has been considerable focus on Boeing in corporate accountability, but the applicable laws are different. In the United States, on the face of it, Boeing may take comfort that corporate accountability has taken a different route, where corporations deemed as being inherently good and law-abiding and any flagrant violation can be dealt with through existing civil laws of vicarious liability. That, certainly, is the view shared by many scholars, but they miss the point that such liability can lead to crippling punitive damages, that are designed to punish the Defendant Company for outrageous conduct that, in other regimes, might be regarded as having the properties of a crime. The award of quantum in punitive damages is a matter for the Jury, and an American Jury can be as capricious as an English one. In any event, they will be concerned to assess how risk management was conducted in drawing their conclusion. No commercial venture is risk-free, and the eternal truth of business management is that directors must balance the chances of an event occurring against the consequences of it occurring. After all, it is generally the high-risk scheme which attracts the investors: such schemes can collapse expensively but, if they succeed, they promise the greatest reward to the company's shareholders. Life can be much more complicated than that, of course, if the director is to keep the shareholders satisfied; that was the priority, at least, of the directors of Ford, and the non-aviation case of Robert Grimshaw makes alarming reading.

Case Study: Grimshaw v Ford

The late 1960s was proving to be a troubled time for the makers of large American motor cars, not least because of a looming global recession and the effect which crude prices had on the demand for motor cars in the United States. Manufacturers had to re-think their business strategies and plan for cars which satisfied the changing demand; failure to get the strategy right would inevitably result in declining market-share and a declining share price for the company. It was, of course, top of the list of the directors' duties to guard against such ills and so, in 1968, Ford approved the concept of their new 'subcompact' family car, the Pinto. It was only after the company was fully committed to production, though, that an engineer pointed out to them that the new design posed a serious risk in a collision, for the fuel installation, crushed into such a confined space, presented a high risk of a fire which could kill or maim the occupants.

The Ford directors did not panic, but calmly assessed the risk, and ordered a cost-benefit analysis. As a result, they were able to establish some forecasts which, admittedly, in the cold light of day, combine to draw a shocking picture of the consequences of making Board decisions without regard to ethical practices:

- The Pinto would probably cause 180 burn deaths, 180 serious burn injuries, and 2,100 burned vehicles, resulting in civil claims, based on the quantum of claims in contemporary cases, of $200,000 per death,$67,000 per Injury, and $700 per vehicle, making a total of $49.5 million.
- On the other hand, the costs for installing safety features would cost some $11 per Pinto, which would give a total bill of some $137.5 million.

As a result, the directors took the management decision to tolerate the risk, which, based on the arguments in the cost-benefit analysis, of course made financial sense. What they apparently did not anticipate, was that their decision also satisfied the legal definition of recklessness, famously held in England by Lord Diplock, that a Defendant in a criminal case is reckless when (1) he does an act which in fact creates an obvious risk that property will be destroyed or damaged; and (2) when he does the act he either has not given any thought to the possibility of there being any such risk or has recognised that there was some risk involved and has nonetheless gone on to do it.[29]

In May 1972, Mrs Gray was at the wheel of her six-month-old Pinto in California, with 13-year old Robert Grimshaw as a passenger. As a result of a rear-end collision, Mrs Gray died from her injuries; Robert Grimshaw survived, but with shocking personal injuries, just as the engineer had predicted. At the trial of his civil claim against Ford,[30] the Jury held against the company and awarded total damages exceeding $128 million, most of which consisted of punitive damages, to reflect the Jury's condemnation of the recklessness of the Board's decision.

Although the quantum of punitive damages in the United States has been reduced dramatically by a succession of judicial decisions in recent years, the core issue so relevant to this paper is clear: the management by directors of risk can be a crucial factor in determining whether the Company's activities are managed or organised in such a way as to cause a person's death, amounting to a gross breach of a relevant duty of care owed by the Company to the deceased. This mirrors the provisions of Section 1 of the Corporate Manslaughter and Corporate Homicide Act 2007 in English Statute Law.

In this way, we can clarify, somewhat, the blurring of the lines which define criminal accountability to punish the wilful and reckless offender, from the notion of Just Culture to promote safety by learning from mistakes.

NOTES

1 See Michaelides-Mateou and Mateou, supra.
2 See Chapter 2.
3 See Chapter 2 and Chapter 3.
4 Chapter 1.
5 Chapter 1.
6 See Australian Government paper *Foreseeability, Standard of Care, Causation and Remoteness of Damage.* www.treasury.gov.au/sites/default/files/2019-03/R2002-001.
7 Lowenfield, A, 1989, Looking Back and Looking Ahead. *American Journal of International Law* 83 (2) (April 1989).
8 ICAO, Airbus A300Bc, EP-IBU, accident in the vicinity of Qeshm Island, Islamic Republic of Iran on 3 July 1988. ICAO Circular 260-AN/154.
9 This summary was extracted from the House of Commons Library Briefing, *Conflict in Ukraine: A Timeline (2014 - Eve of 2022 Invasion).*
10 European Organisation for the Safety of Air Navigation, 2018, Model For A Policy Regarding Criminal Investigation And Prosecution Of Aviation And Railway Incidents And Accidents
11 Regulation (EU) No. 996/2010.
12 See www.justice.gov/opa/pr/belarusian-government-officials-charged-aircraft-piracy-diverting-ryanair-flight-4978-arrest.
13 Aerospace & Defence Industrial Co-operation. See www.epicos.com.
14 Convention on International Civil Aviation, signed 7 December 1944 by 52 States. As of 2013, the Chicago Convention has 191 state parties.
15 11th Edition, 2016.
16 United Nations Convention on the Law of the Sea, signed in 1982, replacing four 1958 treaties. UNCLOS came into force in 1994.

17 SOLAS Chapter V regulation 34-1.
18 Last amended 2022.
19 CAA, 2018, *Information and Guidance on the Circumstances under Which the Direct Costs of a Private Flight May be Shared between up to Six Occupants (Including the Pilot) of an Aircraft.*
20 *R v Murgatroyd*, 2017, Manchester Crown Court (unreported).
21 Trögeler, M, 2010, Criminalisation of Air Accidents and the Creation of a Just Culture, accessed via http://media.leidenuniv.nl/legacy/mildred-tr-366geler-eala-prize.doc%29.pdf.
22 Kind acknowledgments to Ashley!
23 Wygle, B, 1978, *Performance Analysis of the Olympic Airways Takeoff at Athens on August 9, 1978 with an Engine Failure at Rotation*, Boeing. Access via www.scribd.com/document/331020914/1978AUG09-SX-OAA
24 *R v P & O European ferries (Dover) Ltd* [1991] 9 Cr App R 72.
25 The Attorney-General referred questions to the Court of Appeal, which upheld the decision of the Trial Judge. See A-G ref no. 2 of 1999 [2000] 2 Cr App R 207.
26 See Reforming the Law on Involuntary Manslaughter: The Government's Proposals (Home Office, May, 2000, CC NO77828), pp. 30–36).
27 *R v Adomako* (1994) 99 Cr App R 362.
28 *R. v CAV Aerospace Ltd* unreported 31 July 2015 (Central Crim Court).
29 *R v Caldwell* (1982) 1 AER 961.
30 *Grimshaw v. Ford Motor Company* (1981) 119 Cal App 3d 757.

6 Where Does the Theory Part Company with Practice?

OVERVIEW

It is inevitable that any discussion or analysis of 'Just Culture' that ignores the gorilla of gross negligence sitting in the corner will merely perpetuate the problem of criminalisation which, ultimately, would defeat the whole purpose of the policy to protect responsible professionals who are trying to do their job.

Criminal prosecutions that are started following aviation accidents should be pursued when evidence is revealed that the accused individual's conduct demonstrated something more than professional negligence, but how does Just Culture theory get weaved into the web of gross negligence and its impact on criminal accountability and the fairness of the prosecution process?

In this chapter, we examine some of the take-aways revealed in the cases of prosecutions, not only of wilful or recklessness behaviour possessing the *mens rea* and *actus reus* of a crime but also of criminal prosecutions against aviation professionals for acts and omissions that hitherto have been treated solely as civil infractions.

The Just Culture theory is intended to promote continuous learning from previous mistakes and to encourage pilots and other aviation professionals to openly and freely share essential safety related information. The Civil Aviation Authority's (CAA) own approach has been defined as *one in which people are treated fairly and are not punished for actions, omissions or decisions taken by them that are commensurate with their experience and training.* It qualifies this with the proviso that *to maintain or improve aviation safety in cases of, for example, gross negligence, wilful violations and destructive acts, further action may be necessary.*

The distinction between civil liability in negligence and criminal accountability suffers from blurred lines of distinction, with the result that a climate of fear exists between individuals being interviewed and the investigators, resulting in reluctance to disclose critical facts that hinder or even prevent the investigation boards from determining the cause or probable cause of the accident.

DOI: 10.1201/9781003489696-6

At the heart of the problem is the challenge of dealing with evidence obtained in the technical investigation process, which is designed to culminate in safety recommendations, ending up, to paraphrase Ebenezer Scrooge, having every item presented dead against them.

This is where the theory parts company with practice, and we need to confront the practical challenges, before we can start to formulate a solution.

THE HURDLE OF CRIMINALISATION

The very phrase Criminal Negligence appears as a contradiction in terms, but we need to get beyond that and understand its role in the tension between Just Culture and Criminalisation.

In Chapter 1, we were introduced to criminal negligence, when Lord Hailsham famously re-stated Brett J's maxim when he held obiter in *Haughton v Smith*[1] that *the deed does not make a man guilty unless his mind be also guilty.* The crucial issue is that this clarifies the common law and applies also to crimes under statute law, because English statute law falls short in giving guidance on crimes involving negligent behaviour.[2] The excellent reason for this is that, historically, English law did not identify a concept of criminal negligence. In order to dissect and analyse this, therefore, we must seek some guidance elsewhere on just what the State intended; in this context, the American system will serve us well.

For all the faults of the American system as demonstrated in Captain Schröder's case, it tackles the definition of criminal negligence somewhat better. Article 33 of the Criminal Code can be paraphrased to provide a sound starting point with the guidance that an act committed with express intent to bring the desired consequence upon the victim – or with the acknowledgement of the inevitability of such a consequence by the Defendant's action – shall be considered as a crime. This establishes the first of three limbs falling within the definition of *mens rea* as originally developed in Roman Law, with the concept of *dolus directus* (direct intent), where the consequences of an action were both foreseen and desired by the Defendant.

Naturally, it is essential that this eventuality can be rejected in a case of negligence, which conveniently leaves two alternatives to be addressed. In this respect, the range of opinion differs not only as to analysis but even as to definition. The American Criminal Code[3] offers the distinction, and can be paraphrased in that:

- a crime will be committed negligently if the Defendant foresaw the consequence of their action as reckless, that is, if the Defendant saw the risk or inevitability of the consequence in the action, and went ahead with the action, nevertheless, thus allowing the consequence to follow or treating the risk with indifference;
- by comparison, a crime will be committed by negligence if the Defendant either had foreseen the possibility of the risk to the victim and expected without valid reasons that the consequence would be prevented, or had not foreseen the possibility of the consequence but should have done.

By creatively associating common factors from a range of academic opinion in Roman Law, some parallels can be drawn:

1. *Dolus indirectus*, where some collateral consequence of an act falls upon the victim and, although foreseeable as a certainty, the consequence was not specifically desired by the Defendant; nevertheless, he carried on regardless, with full knowledge of the risk.
2. *Dolus eventualis* is defined where the Defendant foresees as a possibility – not necessarily a certainty – the consequence which befell the victim, and still proceeds with the act in question.

The American view eventually runs into logic problems because the analysis of the two as discrete levels of negligence, both in academic arguments and in case law, defy a clear distinction, just as English law struggles with clear analyses of wilfulness and recklessness. Both fall within the category of a crime committed negligently under the American Code, without addressing the crime committed by negligence. On that basis, the American system touches parts that the Roman system cannot reach, namely the Defendant's assumption of the risk – in other words, the consciousness and acceptance of the risk. In essence, therefore, accountability for criminal negligence will depend upon the question as to whether the Defendant's state of mind amounted to recklessness. If the Pilot in Command or the Air Traffic Controller genuinely does not apprehend the risk that a crime could occur as a result of their action, then, at most, they have made an error of judgment, which leaves them blamelessly inadvertent; while such may still leave them accountable in the tort of negligence, it does not incriminate them in a crime.

The aviation professional's accountability necessarily hinges upon their judgment, in which they have the discretion allowed in current law, as it has allowed for generations, to make decisions about what to do and when to do it. How they discharge the burden of their accountability will be determined by their awareness, or conscious assumption, of the risk that will follow. If they hold their belief in the consequences genuinely, then they will have discharged their burden in criminal law – with the aside of Lord Bingham's wisdom that, the more unreasonable the belief, the less likely it is to be accepted as genuine.

Naturally, this entire process is dependent upon gathering the relevant evidence, either to prove the case or to put it in doubt. The Prosecution, therefore, must do the best they can to meet the Evidential Test for a case, but they will want to prove that with a clear, persuasive and rapid decision-making process, making use of all the evidence available.

The tenets in the Chicago Convention define different priorities, of course, with a focus on continuous learning for airline safety. The result is a dynamic tension between the enforcement process of the Sovereign State and the policy of Just Culture which transcends States. The Chicago Convention has tried to compromise with a provision in its general principles that *It is not the purpose of protecting safety information to interfere with the proper administration of justice in States.* But the Convention guards its purpose jealously, by adding "*National laws and regulations protecting safety information should ensure that a balance is struck between the need for the protection of safety information in order to improve aviation safety, and the need for the proper administration of justice". It continues National laws and regulations protecting safety information should prevent its inappropriate use,* and *Providing protection to qualified safety information under specified conditions is part of a State's safety responsibilities.*

EVIDENCE FOR THE PROSECUTION

In Chapter 3, we discussed the cases of *Hoyle v Rogers* in the UK and Captain Schröder in the United States. In *Hoyle v Rogers*,[4] the Court of Appeal upheld the lower court's decision that the AAIB report, containing all the opinions and analysis, could now be admitted in proceedings, giving rise to three particular issues:

1. In accordance with normal rules of evidence, the maker of a report ordered to be admitted under UK law in Part 32 of the Civil Procedure Rules, must be available for cross-examination; otherwise the robustness of assertions of fact or opinion contained in the report would be undermined as unsafe. The effect is that the confidentiality of the investigators contributing to the report may be more cautious about the statements they make, diluting their contribution to air safety.
2. As we have seen above, such opinions which could lead to a Court's finding of negligence in civil proceedings may subsequently underpin a decision to arraign a Defendant on charges of criminal negligence.
3. The downstream consequence is that interviewees in the investigation process may be reluctant to co-operate with the investigating authorities and, despite State regulations compelling their co-operation, would be justified by their privilege against self-incrimination, entitling them to be cautioned before an interview and to have independent legal advice throughout the interview, so that they may have counsel on their right to silence. The further dilution of the value of such investigations to air safety is obvious.

European Regulation No. 376/2014[5] was designed to guard against this. Article 15.2 provides that information derived from occurrence reports shall be used only for the purpose for which it has been collected and that Member States *shall not make available or use the information on occurrences in order to attribute blame or liability... or for any purpose other than the maintenance or improvement of aviation safety.* Article 16 particularly provides an exemption in genuine criminal cases of wilful misconduct or reckless disregard of risk.

Clearly, the case of *Hoyle v Rogers* blurred the use of such a report in applying the rules of evidence in civil proceedings. However the higher Court may have viewed this in their consideration, they could not overlook the danger that you cannot tell a Judge which evidence they can or cannot allow; that is a matter up to them, and if they get it wrong, then you can appeal. All that is needed is the Order of the Judge to admit the report.

At the criminal Trial of Captain Schröder,[6] the Prosecution relied on evidence in the casualty investigation, for which the Defendant had not been cautioned and, when asked if it were a criminal investigation, was told, No. This was subsequently proved to be false, but the evidence was allowed any way, leaving us to draw the conclusion that the State offers protection against self-incrimination, while sanctioning the admissibility of what may amount to a confession gained by unfair means, during the process of an investigation.

The first take-away, therefore, is that the purpose of Just Culture is overridden in domestic jurisdictions to exploit investigation evidence which may assist in prosecutions in the criminalisation process; that is, to pursue cases outside the remit of crimes which require the criminal mental element of wilful or reckless behaviour.

Trögeler[7] identified an underlying factor in the relationship between the accident investigator and the Prosecutor, finding itself in a breakdown of mutual trust, originating with the French investigation of the Mulhouse crash,[8] with the downstream consequence that judicial investigations would now take precedence, enabling them to seize the evidence, and then make it available to the accident investigators. Whether all the evidence is disclosed to the Defence is another matter that may be open to suspicion.

Just Culture owed a great deal to Annex 13 of the Chicago Conference, which declared in its general principles that *the sole purpose of protecting safety information from inappropriate use is to ensure its continued availability so that proper and timely preventive actions can be taken and aviation safety improved.* It then qualified that by tacit permission to depart from this for the purpose of the *proper administration of justice in States.* In this context, it was a pious hope that laws protecting safety information should ensure that a balance be struck between the need for the protection of safety information in order to improve aviation safety and the need for the proper administration of justice, stating that *national laws and regulations protecting safety information should prevent its inappropriate use.*

Article 4.1 of the general principles in Annex 13 provides that exceptions to the protection of safety information should only be granted by national laws and regulations when

a. there is evidence that the occurrence was caused by an act considered, in accordance with the law, to be conduct with intent to cause damage, or conduct with knowledge that damage would probably result, equivalent to reckless conduct, gross negligence or wilful misconduct;
b. an appropriate authority considers that circumstances reasonably indicate that the occurrence may have been caused by conduct with intent to cause damage, or conduct with knowledge that damage would probably result, equivalent to reckless conduct, gross negligence or wilful misconduct; or
c. a review by an appropriate authority determines that the release of the safety information is necessary for the proper administration of justice, and that its release outweighs the adverse domestic and international impact such release may have on the future availability of safety information.

The consequence is that judiciously worded State laws meeting the exceptions in Annex 13, but actually intend to support criminal prosecutions, undermining the theory of Just Culture, so that prosecuting authorities are able to distort the balance of fairness in order to pursue prosecutions in cases that previously had not met the threshold for criminal accountability, which defines the essence of criminalisation.

In Chapter 2, we discussed the case of LATAM Perú Flight 2213. In 2022, an Airbus A320 had been cleared for take-off on a domestic flight from Jorge Chávez International Airport in Lima. The Pilot in Command had not received any warning of a runway incursion, and only at the last moment did he observe a fire truck dead in his path. He exercised remarkable skill in endeavouring to mitigate the risk, but collided with the truck, when the starboard engine separated from the wing and the starboard landing gear collapsed. The aircraft then slewed across the runway on fire, and came to rest 8,200 feet down the runway.

Far from being praised for their textbook crisis response and flying skills avoiding an imminent disaster, the pilots of the aircraft were arrested and held in custody for four days. Having acted with great professionalism they were subjected to the extreme stress of being held in custody in anticipation of charges which, no doubt, would have included the homicide of the firefighters who had been killed in the truck, and were denied the psychological medical care that their situation demanded. The prosecuting authorities gave no thought to Annex 13 of the Chicago Convention, and it was nearly a year before the Peruvian Accident Investigation Commission issued their report. The investigation concluded that a lack of joint planning and poor coordination of the fire rescue exercise, coupled with a failure to use standardised communication and phraseology, were the probable causes, naming ten contributing factors, none of which attached responsibility to the flight crew. It has to be said, that the findings cast some accountability on the Air Traffic Controllers, but no prosecutions have been launched against any aviation professional in this case. The suspicion must also be aroused that Annex 19 was not followed, establishing a safety management system through the consolidation and reorganisation of Standards and Recommended Practices, particularly concerning the appropriate use and protection of safety information and their sources; a criticism which could be applied in other cases discussed in this book, involving the investigation and prosecution of aviation professionals.

The UK prosecuting authority, the Crown Prosecution Service, seeks to justify a decision to prosecute on decisions made by a CPS officer drawing their own answers to the Full Code Test. Chapter 4 examined this in some detail, but, even if it passes the Evidential Stage to meet a likely conviction on the defined offence, the Public Interest questions can be examined in the following questions:

- How serious is the offence committed?
- What is the level of culpability of the suspect?
- What are the circumstances of and the harm caused to the victim?
- What was the suspect's age and maturity at the time of the offence?
- What is the impact on the community?
- Is prosecution a proportionate response?
- Do sources of information require protecting?

Notably absent is any reference to the theory defined in Just Culture. The prosecuting authority has no obligation to make practical decisions that meet the demands of continuous learning and understanding the role of mistakes in patterns of behaviour; as a result, it has no guidance on the role of technical expertise in deciding whether or not the professional's risk assessment and behaviour is acceptable within the norms of their professional duty. It simply may draw conclusions on the tenet that reckless or wilful misconduct will not be tolerated. While it is noted that CAA's approach has been defined as one in which people are treated fairly and are not punished for actions, omissions or decisions taken by them that are commensurate with their experience and training; nevertheless, the decision to prosecute involves a discretion that raises uncertainty and, therefore, discourages the individual from reposing their trust in the system to establish fairness.

The solution would certainly require clear definitions of recklessness and misconduct. Such issues are addressed by International Civil Aviation Organization (ICAO) guidance[9] in defining misconduct as *the non-compliance by staff members, through acts or omissions, with their obligations under the Chicago Convention, the ICAO Service Code, Staff Rules, Personnel and Administrative Instructions and other relevant and administrative texts in force.* It adds that misconduct *is also the non-observance of standards of conduct expected from an international civil servant.*

This does not help the Prosecutor very much in the determination as to whether the evidence meets the threshold requirements for the *mens rea* and *actus reus* of an offence. As far back as 1982,[10] the House of Lords addressed the meaning of recklessness when it formed part of the *mens rea* of a crime, when it held that the word had not acquired any special meaning as a legal term of art, but bore its popular meaning of carelessness, regardless or heedless of the possible consequences of one's acts. It meant that a Defendant could be held to be reckless even though they never actually foresaw the risk occurring, if the Jury concluded that the risk would have been obvious to the ordinary and prudent bystander. This was unsatisfactory and evolved with the case of *R v G* in 2004,[11] in which it was held that a person acts recklessly with respect to a circumstance when they are aware that a risk exists or will exist, in so far as they should only be regarded as acting recklessly where the risk would have been obvious to them and they had given thought to it. The mental element is obvious and, crucially, its proof must be subjective.

In this situation, the question arises as to where the factor of human error crosses the border-line to end up in criminalisation, where the Prosecutor can identify a realistic chance of success with a case of criminal negligence, regardless of the Defendant's state of mind. The case of *Adomako*[12] set the precedent, taking the principles of civil liability in negligence one step further, in which the Jury had to consider whether the breach was so grossly negligent that it should be seen as criminal.[13] The awful truth is that it does not matter whether the Jury analysed the expert and eye-witness evidence correctly, because their deliberations will not be made known to the Court; only the verdict will be recorded.

The consequence leads to some critical analysis of the general principles in Annex 13 in cases which the Jury either fails to understand or ignores critical evidence on the Defendant's risk management decisions – evidence which may be highly complex – and they are, rather, more tempted to place reliance on the Prosecution's assertion of reckless conduct, gross negligence or wilful misconduct. The continuous learning process in Just Culture will be seriously at risk in the future, when professionals will be fearful of the dangers of co-operating with the process and have the consequences of their mistakes presented dead against them. They will rely, instead, on the advice of their lawyer to decline to answer a question in reliance on their privilege against self-incrimination. Worse still, the human nature of self-preservation may induce them to assess the strengths or weaknesses in their position and colour their report or evidence accordingly; such an outcome can conceal serious weaknesses that investigators might otherwise reveal and address to improve flight safety.

JUSTICE OR CRIMINALISATION?

The second take-away takes us deeper into the criminalisation process. In Chapter 2, we examined the nature of criminalisation, the process by which crimes are created which lead to the prosecution of offenders that previously were not considered crimes by act or omission, but which have been made crimes by Society notwithstanding the human rights of the offender.

It is all down to how Society prioritises rights and obligations. Moral standards regulate right and wrong conduct, which underpin the concept of Justice, that is, the classification of moral rules which contain the essentials of human well-being. Society's moral standards constantly evolve, responding to how it prioritises its threats and values. Hence, Society may prioritise key issues which it perceives present some common danger; and, as we have seen repeatedly, what Society wants, Society gets.

The mischief, which broods at the very heart of the question, demands an understanding of just how the concept of criminal justice has evolved in the business of accountability. Such accountability logically should be defined by a single principle against which we judge all actions. In civil aviation, this mischief is all the more shocking, because it is new. Once again, we can get an understanding of its evolution through the somewhat longer history of the accountability of the shipmaster, and by applied logic we can form an opinion about the foundation upon which it is based. There are immediate and obvious parallels, given the Master's responsibility for the vessel and the safety of life under the laws of the Flag State, in balance with (sometimes in conflict with) the commercial responsibility to their employer, subject to the duty to observe the laws of the States into whose jurisdiction they operate. Yet, the possibility fades for a definition of a single principle of criminal justice, that can be applied consistently across the many jurisdictions into which the Master or Pilot in Command may get into trouble. It is the evolutionary process of criminal accountability across these jurisdictions, defined and applied by moral rules that differ according to the normative ethics of changing societies, which has forged the process which the global community identifies as the phenomenon of criminalisation.

Case Study: Captain Chawla of the *Hebei Spirit*

At about 07.30 local time on 7 December 2007, near South Korea's Port of Daesan, a crane barge owned by Samsung Heavy Industries, that had been involved in a towage operation in which a tug had lost control after the line parted in heavy seas, collided with the *Hebei Spirit*, carrying 260,000 tonnes of crude oil. Although no casualties were reported, the collision punctured three of the five tanks aboard the *Hebei Spirit* and resulted in the leaking of some 10,800 tonnes of crude.

The spill occurred near one of South Korea's most beautiful and popular beaches, contaminating one of Asia's largest wetland areas. It was disastrous for wildlife because these areas were relied upon by migratory birds as well as resident wildfowl and marine life. As a national maritime park, with a host of fish farms, the region provided the living for hundreds of Korean families. Now these resources were annihilated, under the glare of the world's media, which compounded the humiliation of the people and their government.

The South Korean government declared a state of disaster in the region, with a clean-up cost estimated at US$330 million, involving 30 aircraft and 327 vessels. But that was not enough to satisfy the anger of the People, who were looking for someone to blame and punish. The political consequences of inaction would rapidly manifest themselves at the ballot box, and the government scrambled to find that someone, who was, of course, the Master and his Chief Mate.

South Korean criminal procedure shares features both of the adversarial and inquisitorial systems, having been founded on a merger of American and German criminal procedures. It was not until February 2008 that a limited system of jury trials was adopted for criminal cases and environmental cases, and all questions of law and fact are determined by judges.

On reflection, it was, perhaps, unsurprising that normative ethics of South Korean society demanded that the State should react with criminal prosecutions against those whom it believes to be accountable. Naturally, the evidence must first be examined in order to establish the facts of the case which, no doubt, would establish a solid foundation upon which to prosecute the Accused. But the trial of Captain Jasprit Chawla, an Indian national, pre-dated the casualty report dramatically. In a press statement released by the Prosecutor's Office in January 2008 after the indictments had been signed, it was announced that the decision had been taken to prosecute Captain Chawla as well as his first officer of the *Hebei Spirit* and the skippers of the barge and two tug boats that towed it, all for failure to exercise due caution:

> Captain Chawla (of Hebei Spirit) overlooked the possibility of collision and carelessly assumed that the tug boat fleet would safely pass by a distance of about 280 metres (918 ft).

The South Korean criminal legal system distinguishes between serious offences and less serious offences, and crimes of a violent nature are distinguished from property-related crimes. There was no element of dishonesty in the case against Captain Chawla, mitigating the seriousness of the case against him even further. In the event, the Trial Judgment, in June 2008, cleared Captain Chawla and his Chief Mate of all charges, although the skippers of the tugs were jailed for marine pollution offences – one of whom was also convicted of a serious offence of dishonesty, for falsifying navigation records.

The State Prosecutors could not possibly accept this decision and lodged an appeal, which would require a re-hearing of the case against Captain Chawla and his Chief Mate. It was now abundantly clear that the State's Central Maritime Safety Tribunal would have to deliver its final adjudication well in advance of the appeal hearing. As it happened, it took just under a year for that to happen.[14] Responsible Examiner Jong Eui Kim concluded that the collision was caused when the crane barge ensemble lost its towing ability in bad weather, yet still continued to navigate without taking any safety measures, such as warning the other vessels nearby or performing emergency anchoring. Despite the fact that the *Hebei Spirit* was anchored in an area frequented by navigating vessels, it was found that her command management was negligent (in the understanding of civil liability which we have previously discussed) in performing its duty and was idle in handling the situation, whereby it failed to take early and active preventative measures. In addition, it was averred the fact that the tanker's main engine was not operable, amounted to negligence in engine readiness, in a situation where there was a risk of collision, resulting in the failure to take preventative measures. The examiner concluded that the cause of South Korea's worst-ever pollution event was the crane's jib puncturing the cargo tanks of the *Hebei Spirit*, resulting in the cargo spilling into the sea. The report followed with a conclusion that the pollution event had been exacerbated due to *inappropriate emergency measures taken by Hebei Spirit after the collision occurred, which increased the speed at which the oil spilt.*

However argued, flawlessly or otherwise, the Tribunal's casualty report was finally signed on 4 December 2008 – just six days short of the date set for the appeal hearing, when the Court reversed the decision of the lower Court and convicted Captain Chawla, who was sentenced to 18 months imprisonment, not for the physical spill, but on two charges alleging property-related offences of criminal damage – damage to his own ship, that is. As regards his actual misconduct, the Court held that Captain Chawla's criminal accountability rested on a number of key mistakes, although only limited fault attached to his acts or omissions before the crane barge actually hit the vessel. Before the collision, he should have gone full astern to drag anchor to prevent the collision with the drifting crane barge. Then after the collision, he should not have pumped inert gas into the cargo holds, which had the effect of increasing the spillage when the explosive risk was low. Additionally, he should have shifted ballast in order to create a sufficient list to take the cargo away from the hole in the damaged hull, which would have prevented the oil spill.

Society was appeased; Agence France-Presse reported that about 100 residents of the area affected by the spill clapped outside court after the judges issued their Judgment. One said laconically,

> We are satisfied with the verdict.[15]

But the global maritime community was appalled. Following the conviction, International Transport Workers Federation maritime coordinator Stephen Cotton articulated the industry's outrage in a comment to Lloyd's List:

> This is not justice. It's not even something close. What we have seen today is scapegoating, criminalisation and a refusal to consider the wider body of evidence that calls into question the propriety of the court. This decision is incomprehensibly vindictive and will impact on all professional mariners…[16]

In April 2009, the Supreme Court heard the Defendants' appeals, and accepted the appeal in relation to the somewhat bizarre charge of the wilful destruction of the *Hebei Spirit*, for which they had received prison sentences. It had been a miscarriage of justice, of course, but it is the law, and the case followed the laws and procedures of

South Korea, a sovereign State. It is this tension between justice and fairness which has brought the topic of criminalisation to the top of the global agenda.

APPLYING THIS CASE TO THAT OF CAPTAIN CHAFIK AL GHARBI

Just a month separated Captain Chawla's case from the prosecution of Captain Chafik Al Gharbi of Tuninter Flight 1153. In March 2009, an Italian court sentenced Chafik Garbi to ten years in prison for manslaughter. Prosecutors submitted that after the plane's engines stopped functioning after he succumbed to panic, started praying and failed to follow emergency procedures, and that he could possibly have reached Falcone-Borsellino Airport. Six others, including the co-pilot as well as the chief operating officer of Tuninter Airlines, were sentenced to between eight and ten years, which were subject of appeals.

The International Federation of Airline Pilots Associations was as critical of the conviction, as the International Transport Workers Federation had been in the case of Captain Chawla. At their 64th conference, IFALPA unanimously expressed their dismay when informed that, once again, the compulsion to apportion blame had outweighed the greater need to improve the safety of air transport. They submitted that, in all the circumstances, the prosecution was misconceived given the evidence in the case, with a harmful effect on air safety in the future. Going somewhat further, IFALPA pointed out that this appeared to perpetuate a dangerous trend in Italy of its pursuit of accident related prosecutions; and called on the Italian government to take immediate steps to amend the laws which, IFALPA observed, prejudiced air safety and the downstream consequence of the safety of the travelling public.[17] Of course, IFALPA could go no further, neither could any organ in the international community, for to do so would compromise Italy's sovereignty.

SUMMING UP THE ROLE OF SOCIETY

We have seen how Society's normative ethics evolve in response to internal change and external stimuli and, with it, the concept of a risk which might threaten the security of that society has evolved. Indeed, Ericson and Carriere have defined this in terms of what they label a 'risk society,' in which society has become more concerned with public safety than with the old preoccupations with social inequality.[18] It is a very logical consequence, therefore, that the concerns of the twenty-first-century risk influence the dynamic on the definition of a moral wrong in the context of normative ethics; but that does not necessarily make it fair. Perhaps it is unsurprising, then, that the theory of Just Culture finds itself parting company with practice, because society places a high value on continuous learning for the future of air safety; but it actually places a higher value on accountability and blame. Democratic systems have no choice but to meet the demands of society's values; after all, that was the basis upon which they won society's vote into power, and society can just as easily vote them out of office at the next election.

This question of blame was addressed with great perception by Victor Tadros in his paper *The Homicide Ladder* in the *Modern Law Review*,[19] with his observation of human nature, that, when a sudden death takes place following an accident, society will look around for somebody to blame. In criminal prosecutions in maritime cases, this has become a worrying trend of finding a scapegoat to satisfy public demand following a fatal accident or an environmental disaster; in civil aviation accidents, the same search for a scapegoat could be identified, except that, in such cases, prosecutions are rarer because the pilot in command has died. The words of Lord Atkin in *Andrews v DPP* which we met in Chapter 1 are worth quoting in a little more detail:

> Of all crimes manslaughter appears to afford most difficulties of definition, for it concerns homicide in so many and so varying conditions… the law … recognises murder on the one hand based mainly, though not exclusively, on an intention to kill, and manslaughter on the other hand, based mainly, but not exclusively, on the absence of intent to kill, but with the presence of an element of 'unlawfulness' which is the elusive factor.[20]

The European Court of Human Rights had no qualms, however, in upholding the decision of the Spanish Court to hold Captain Mangouras as the scapegoat in a serious environmental spill. In its written Judgment, it upheld the lower Court's bail decision in meeting, not the interests of justice but *the growing and legitimate concern both in Europe and internationally in relation to environmental offences and the tendency to use criminal law as a means of enforcing the environmental obligations imposed by European and international law.*[21]

In consequence, the Court served to perpetuate criminalisation in the professional environment which made demands upon a professional such as a shipmaster or pilot in command. The Court went further and approved the relevance of social anxieties about consequential damage beyond that of a fatal accident, because that was what the new Risk Society was demanding.

Sidney Dekker took an analysis of the Risk Society a step further in asking the question: *Why do we blame?* He posed the theory that Society demands a scapegoat, possibly, as the easy way out for justifying its illusion that Society has control over its increasingly technological risks.[22] The Brave New World of the twenty-first century is developing technologies that, unquestionably, leave the understanding of people fare behind, and put them out of human control, which creates anxiety, so it needs to create an illusion to comfort itself, with somebody who must be accountable, irrespective, perhaps, of fault. The UK Parliament has, itself, acknowledged Society's anxiety with new technology, specifically addressing the example of artificial intelligence:

> It's hard to guess how AI will affect our lives in the future, although the impact so far has already been significant. While AI has immense potential to aid society, such as improving healthcare outcomes, it could also create social and ethical harms including greater inequality.[23]

The only control that Society has in the ever-increasingly complex world of technology is to attribute blame when something goes wrong, even if that means finding a scapegoat to soothe its fear. The process for that already exists, in the Court system, in which Judges must follow sentencing guidelines according to the level of culpability and harm suffered in a case.[24]

It may, indeed, be argued that this is the deep chasm that divides Just Culture from the judicial process and reveals where the theory parts from practice. The Just Culture process has its focus on finding the causes, in other words, it asks the question: *What was responsible?* The investigation is designed to determine the engineered and procedural circumstances for the crash, in order to avoid mistakes in the future. The legal process, however, asks the question: *Who was responsible?* It puts a person in the dock to determine whether the evidence shows beyond reasonable doubt whether they committed the criminal offence which Society alleges against them. Unless the answer to the first question leads inexorably to incrimination in the second, the two questions are so different in nature as to be incapable of reconciliation.

Society's demand for accountability will not be denied, however. As a result, there must be a solution in compromise, balancing safety and accountability. That is the fundamental policy of Just Culture; how States grapple with that in their systems is another matter.

CRIMINALISATION AND SOVEREIGNTY

Lurking mischievously behind the second take-away is a third. We have seen how international condemnation flowed from the conviction of Captain Chafiq. But, just as Captain Chawla's conviction was condemned by a host of international voices, they had no teeth with which they could force a reversal of the decisions.[25] You cannot interfere in the legal process of another State. The United Nations Charter, Article 2 of which defies any misunderstanding:

1. The Organization is based on the principle of the sovereign equality of all its Members.
2. All Members, in order to ensure to all of them the rights and benefits resulting from membership, shall fulfil in good faith the obligations assumed by them in accordance with the present Charter.
3. All Members shall settle their international disputes by peaceful means in such a manner that international peace and security, and justice, are not endangered.
4. All Members shall refrain in their international relations from the threat or use of force against the territorial integrity or political independence of any state, or in any other manner inconsistent with the Purposes of the United Nations.

By the same token, the defaulting State can take comfort from the fact that the United Nations has no teeth with which to enforce this. The position taken by Russia in the case of the downing of MH17 is a pertinent example and should have proved a warning sign to the West of the dangers of turning a blind eye to a State's criminal ambitions under international law.

For this reason, the noble ideal of State Sovereignty has become an obstacle to the concept of fairness in the way in which pilots are treated – or mistreated – in different systems, which, at least on the face of it, are obliged to follow international polices as defined in treaties such as the Chicago Convention.

On the face of it, Captain Chafiq's conviction would appear compelling evidence of the criminalisation of the pilot, as part of an evolving process of risk management within Sovereign States, applying the basics of responsibility to the normative ethics of the Risk Society, and defining the law in order to meet the demand for convictions, even dispensing with the necessity of establishing beyond reasonable doubt that the Accused had possessed the necessary *mens rea* for the offence in question. In the process, the pilot's human rights had been ignored.

As far back as the 1980s, Raymond Michalowski harnessed two very academic theories involving criminal justice to society's control of harmful behaviour, and ending up with a generalised concept of criminalisation by which the individual's behaviour or conduct is transformed from what civil law would call something more than an error of judgment, into a criminal offence.[26] This truly defines the foundation of criminalisation, and reveals it as an emerging problem, insofar as it has been precipitated by the normative ethics of society which has led to an evolving pantheon of crimes that have found their way onto the statute books of sovereign States, and nobody else can interfere in the process. Then it takes a step further, and finds common ground with other sovereign States, resulting in Conventions which set new international boundaries in criminal accountability for the violation of laws that may or may not demand a mental element of Guilt. It is up to the individual State as to how zealously it implements that in its sovereign domestic legislation.

While this is going on, the human element finds itself abandoned by the law-makers. After all, theories are all well and good as bases for justifying new law, but in reality, it is their effect on the human element which weighs upon the pilot in command, because they define and reset the ethical application of the criminal law to their professional judgment, which is where the human element lies. As such, these new laws and processes must be applied subjectively to the pilot, because their judgment is subjective. It is not objective, which society might prefer as a tidier way to justify a conviction on a legal definition, and sometimes that definition has to be adjusted to meet the criteria for a crime; the evidence for which is only too apparent in the evolution of the relatively new crime of Criminal Negligence, which finds itself defined in different jurisdictions as their laws are applied. In many such cases, the crime arises out of liability for violating operating procedures, as a result of the risk assessment which forms part of the Pilot's discretion; the consequences may result from pilot error, hardly reaching the level of civil negligence let alone displaying the mental element of a crime. Notwithstanding the unfairness of such a process, the fact remains that it is still the law, and the pilot inevitably must weigh the balance of such risks while managing the crisis in hand; the human nature of self-preservation is a very strong one.

So, we are back at the question which we posed in Chapter 4: *Who gets to draw the line?* The answer lies in the laws and processes which devolve on the Prosecutor to decide whether a person should be charged and, if so, what that offence should be; that is a matter of sovereign jurisdiction, extended by the law-makers, in which no other State can interfere. Naturally, the Prosecutor's decision is entirely dependent on the evidence available to them, upon which there are strict limits to available resources. Paragraph 1 of Article 14 of EU Regulation 996/2010 states clearly:

> The following records shall not be made available or used for purposes other than safety investigation:

a. all statements taken from persons by the safety investigation authority in the course of the safety investigation;
b. records revealing the identity of persons who have given evidence in the context of the safety investigation;
c. information collected by the safety investigation authority which is of a particularly sensitive and personal nature, including information concerning the health of individuals;
d. material subsequently produced during the course of the investigation such as notes, drafts, opinions written by the investigators, opinions expressed in the analysis of information, including flight recorder information;
e. information and evidence provided by investigators from other Fl... countries in accordance with the international standards and recommended practices, where so requested by their safety investigation authority;
f. drafts of preliminary or final reports or interim statements;
g. cockpit voice and image recordings and their transcripts, as well as voice recordings inside air traffic control units, ensuring also that information not relevant to the safety investigation, particularly information with a bearing on personal privacy, shall be appropriately protected, without prejudice to paragraph 3.

Paragraph 2 takes this further:

> The following records shall not be made available or used for purposes other than safety investigation, or other purposes aiming at the improvement of aviation safety:
>
> a. all communications between persons having been involved in the operation of the aircraft;
> b. written or electronic recordings and transcriptions of recordings from air traffic control units, including reports and results made for internal purposes;
> c. covering letters for the transmission of safety recommendations from the safety investigation authority to the addressee, where so requested by the safety investigation authority issuing the recommendation;
> d. occurrence reports filed under Directive 2003/42/EC.
>
> Flight data recorder recordings shall not be made available or used for purposes other than those of the safety investigation, airworthiness or maintenance purposes, except when such records are de-identified or disclosed under secure procedures.

But State sovereignty will not be denied. Paragraph 3 states:

> Notwithstanding paragraphs 1 and 2, the administration of justice or the authority competent to decide on the disclosure of records according to national law may decide that the benefits of the disclosure of the records referred to in paragraphs 1 and 2 for any other purposes permitted by law outweigh the adverse domestic and international impact that such action may have on that or any future safety investigation.

This provides the justification for the Court to decide on the disclosure of the report in *Hoyle v Rogers*[27]which was subsequently clarified in the 2019 case of the *Ocean Perfect* in which the Commercial Court ruled that accident investigation reports are inadmissible as evidence as to liability without the express permission of the Court.[28]

Given this permissive authority, a greater mischief lies in the abuse of the power, for the purpose of prosecuting the offender. Trögeler pointed out that statements clearly intended to be protected by EU 996/2010 were increasingly being misused by their admission in evidence in judicial proceedings, and the author must agree. The root cause of the misuse can only lie in the discretionary power of the Prosecutor to draw the line, a concept with which we are now familiar, and then the mischief is perpetuated by the Judge. Certainly, it is open to the Defence team to cross-examine the expert, in order to minimise the risk of misunderstanding of their evidence, but the very fact of admitting the evidence gives rise to the risk of injustice as a result of ignoring the imperatives in Article 14.1 of 996/2010.

The author is not certain about the meaning and extent of Trögeler's proposed solution of equipping accident investigators with legal knowledge of law or procedure, in order to identify the risk of compromising the liability of aviation professionals in the text which they choose in their report. The danger lies in encouraging the expert to make value judgments about the law at all. An Expert Witness has a strict role in the Trial process. They can be anyone with knowledge or experience of a particular field or discipline beyond that to be expected of a layman. The Expert Witness's duty is to give to the Court or tribunal an impartial opinion on particular aspects of matters within his expertise which are in dispute, not to express opinions of law, for which they are not qualified.

In England and Wales and many other jurisdictions, the Court must give permission for an expert witness to give evidence. Part 19.2 of the UK's Criminal Procedure Rules defines the expert's duty to the court:

1. An expert must help the court to achieve the overriding objective[29] –
   a. by giving opinion which is –
      i. objective and unbiased, and
      ii. within the expert's area or areas of expertise; and
   b. by actively assisting the court in fulfilling its duty of case management under rule 3.2,[30] in particular by –
      iii. complying with directions made by the court, and
      iv. at once informing the court of any significant failure (by the expert or another) to take any step required by such a direction.

The Rules emphasise that their duty overrides any obligation to the person from whom the expert receives instructions or by whom the expert is paid.

To summarise, an investigator involved in providing the report will be acting as an expert witness, making findings of relevant facts in their area of expertise on the subject matter of the case, and drawing conclusions that will be open to scrutiny by both Prosecution and Defence; it matters not who pays their fee. The Court, therefore, will rely on their impartial, expert assessment of the facts; it is a dangerous thing to express or imply a statement which can be misread as a legal opinion or conclusion in their report, and cross-examination on such things can lead to doubt over the value of the expert's evidence. Indeed, expert witnesses, themselves, will likely be very wary of drafting their report in a way that may lead to some misconstruction, as their oral evidence under cross-examination may lead to damaging conclusions on their professional credibility. While the expert no doubt will benefit from training in the art and science of compiling a report, such training must be solely in terms of technical drafting; no excursion into liability can be tolerated.

The report, naturally, will be highly influential in persuading the Prosecutor as to whether the line has been crossed and the elements of a criminal offence can be identified as a result, in serious crimes arising out of fatal accidents, whose elements are defined by Statute. So, for example, Section 1 Corporate Manslaughter and Corporate Homicide Act 2007 states: *An organisation to which this section applies is guilty of an offence if the way in which its activities are managed or organised (a) causes a person's death, and (b) amounts to a gross breach of a relevant duty of care owed by the organisation to the deceased.* The Defendant will be arraigned on an indictment under this provision, and the Court's only task in interpretation is to determine what Parliament intended in the wording of the act. The Prosecutor's duties and powers are strictly limited to this interpretation.

The State's sovereign laws and regulations will also determine, however, what offences have been breached besides those defined in a Statute. Parliament alone can make laws which hold an individual accountable, but Parliament has the sovereign jurisdiction to delegate powers to its administrative departments, be they a government department or the Privy Council, which are defined in the Statute only by their limitation. Hence, Part III of the Civil Aviation Act 1982 makes provision for the regulation of civil aviation in the United Kingdom and confers subordinate powers on the Department of Transport to make laws which regulate air navigation, while deregulating measures from the government's General Aviation Red Tape Challenge, but addressing specific offences prescribing penalties, by way of statutory Regulation. Hence, Article 240 of the statutory instrument Air Navigation Order 2016 states: *A person must not recklessly or negligently act in a manner likely to endanger an aircraft, or any person in an aircraft.* The Order does not define any of the elements of *mens rea* or *actus reus*, so it is up to the Prosecutor to decide what is reckless or negligent or likely to endanger an aircraft. Strict liability laws contain obligations which, if violated, will constitute an offence, without a mental element, and will attract the penalties defined in Chapter 4 of the Order. While international air regulation defines many such laws, it is the State's sovereign jurisdiction to enact laws which will lead to criminal accountability. The Prosecutor, therefore, has the duties and powers to draw the line; such powers are devolved by the Sovereign power, and thus it is they who must draw the line between honest mistake and criminal behaviour. Such things are findings in law, which can only be drawn after expert evidence has been delivered on facts contained in the accident report, which would have the probative value on which a prosecution can rely.

This delivers some logic to the provision in Regulation 996/2010 that *the administration of justice or the authority competent to decide on the disclosure of records according to national law may decide that the benefits of the disclosure of the records for any other purposes permitted by law outweigh the adverse domestic and international impact that such action may have on that or any future safety investigation.* Such sovereign powers, however, lack the quality of certainty in the decision to prosecute, which leads to the fear that the aviation professional will not be afforded the protection of a Just Culture to put safety over punishment for something which a prosecutor deemed to have crossed the line to criminalisation. At the heart of this is the risk that their professional understanding of acceptable behaviour, in the highly skilled task of aviation risk management, may be fundamentally different to that of the Prosecutor drawing the line and applying it to the Code for Prosecutors according to standard guidelines, leading to the individual's fear of unreasonable prosecution. The individual does not – or at least should not – seek immunity from prosecution but, rather, seeks some foundation of trust on which they can expect fairness, and that demands some clear definition about where the line is drawn between what is acceptable behaviour and what is a criminal offence, according to the Risk Society by which criminalisation has evolved.

The professional is also a member of the Risk Society and accepts the fairness of the Just Culture, in which they will be held accountable for professionally unacceptable violations of aviation rules, while being encouraged to co-operate in the process of gathering safety-related information upon which expert recommendations can be made in the future. This is the balance of fairness upon which the policy of Just Culture is based. The use of criminal accountability by the sovereign power in abuse of this balance puts the critical element of trust in jeopardy, in which case Just Culture loses its foundation.

Against this, we must also understand the difficulty faced by Prosecutors in the modern context of criminalisation. In the preface to the 2009 edition of *Archbold*,[31] the Bible of English criminal law, the author lamented the need to repeat the recurring theme that there is far too much criminal legislation: *The state of the criminal statute book is a disgrace*. But the rate of criminalisation has only accelerated since then, increasing the burden on Prosecutors who must decide whether an action falls on the side of acceptable or unacceptable behaviour in the highly technical environment of civil aviation in which the professional must make command decisions, often under stress. They must apply the facts to their understanding of what is meant by negligent or reckless behaviour, which at best is vague and must be decided, not only on an industry-specific basis but also on a case-by-case basis, demanding the wisdom of Solomon to determine whether the behaviour in the case under consideration crossed the line between acceptability and criminal accountability, according to the statute. They can only do so by close scrutiny of the evidence, which must serve to prove every element of the crime beyond reasonable doubt, and without the domain expertise of the aviation professional, they are reliant on expert guidance. Most certainly must they address the Evidential Test without any hint of hindsight bias, which would distort an evaluation of the case and lead to the unsafe conclusion that the consequences could have been foreseen in this particular case, so putting in doubt just how accountable the individual was in responding to the emerging risks.

Having passed the Evidential Test, the Prosecutor must then consider whether criminal proceedings are in the public interest. As we established in Chapter 1, there is no definitive guidance as to when it may or may not be in the public interest to prosecute. Each decision will be on the individual facts of the case, and even if the Evidential and Public Interest Tests are served in the Prosecutor's mind, they must still consider whether other factors are present, and apply them on a case-specific basis, such as whether, and to what extent, the emerging circumstances were beyond the individual's control and whether the consequences were reasonably foreseeable in the individual's risk treatment, whether the offence which the Prosecutor is contemplating was the result of a genuine mistake or misunderstanding, as well as other relevant matters to be taken into consideration, not the least being the reliability of the prosecution in the face of non-co-operation or reluctance of witnesses, particularly among potential expert witnesses who may be reluctant to defend their conclusion that, given all the circumstances, they would have acted any differently.

Finally, they would have to consider whether the proceedings would be an abuse of process, particularly given the prioritisation of criminal proceedings over the demands of Just Culture in meeting the balance of fairness.

That being said, the decision on where the line is drawn, followed by the decision to prosecute and the subsequent Court process, are all matters which Parliament has delegated to the executive, but they set their stamp of sovereignty on it, nevertheless. There can be no external interference in the process and, if, at the end of the proceedings, the Defendant feels that there has been a miscarriage of justice, they can seek to appeal. They are on their own.

NOTES

1 *Haughton v Smith* [1975] AC 467.
2 This statement must be qualified with the advent of the Corporate Manslaughter and Corporate Homicide Act 2007; more of which, below.
3 Criminal Negligence, Article 33.1.
4 See Chapter 3.
5 Amended by Regulation 2018/1139.
6 See Chapter 3.
7 Trögeler M, 2010, *Criminalisation of Aviation Accidents: Creation of a Just Culture: Balancing Aviation Safety and the Proper Administration of Justice*, Leiden University Law School, Leiden.
8 Chapter 3.
9 See www.icao.int/secretariat/SecretaryGeneral/Ethics.
10 *R v Lawrence* [1982] AC 510.
11 *R v G* [2004] 1 AC 1034.
12 See Chapter 1.
13 *R v Adomako* [1995] 1 AC 171.
14 Central Maritime Safety Tribunal, Decision Junghaeshim No. 2008-26, Marine pollution caused by the collision of "Samsung No. 1" (barge), towed by tugboats "Samsung T-5" and "Samho T-3," and M/V "Hebei Spirit" (oil tanker), Notice of Decision on 4 December 2008.
15 See AFP newswire report, Court jails tanker officers over S Korea's worst oil spill, 10 December 2008; See www.afp/article/ALeqM5i762frrt2NK0zY7cuE7p6WxxRJiw.
16 Lloyd's List, 13 December 2008, Informa plc, London.
17 See www.eurocockpit.eu/sites/default/files/Tuninter.
18 Ericson, R and K Carriere, 1994, The Fragmentation of Criminology, in D Nelken (ed), *The Futures of Criminology*, Sage, London.
19 V. Tadros, (2006) 69(4) MLR 601–618, at p. 601.
20 *Andrews v DPP* [1937] 2 All ER 552, at pp. 554–555.
21 *Mangouras v Spain* (12050/04). See Chapter 3.
22 Dekker, S, 2012, *Just Culture: Balancing Safety and Accountability*, CRC Press, Boca Raton, FL
23 Bhatnagar, A and D Gajjar, 2024, How Is Artificial Intelligence Affecting Society? https://post.parliament.uk.
24 See Chapter 1.

25 In April 2012, the Appeal Court reduced the sentences of seven Tunisian airline personnel and, following a second appeal, Captain Chafik Gharbi received six years and eight months.

26 Michalowski R, 1985, *Order, Law and Crime: An Introduction to Criminology*, Random House, New York, p. 6: *the process by which behaviors* [sic] *and individuals are transformed into crime and criminals.*

27 See Chapter 3.

28 *Ocean Prefect Shipping Limited v. Dampskibsselskabet Norden AS, the Ocean Prefect* [2019] EWHC 3368 (Comm).

29 1.1 (1) The overriding objective of this procedural code is that criminal cases be dealt with justly.

30 3.2 (1) The court must further the overriding objective by actively managing the case.

31 Archbold, 2009, *Criminal Pleading, Evidence and Practice*, Sweet & Maxwell and Thomson Reuters, London.

7 How Can We Offer a Solution?

OVERVIEW

It is very apparent that States will not abandon the concept of criminalisation, but without that sense of fairness which underpins a Just Culture, the individual or company will have no confidence that their rights and interests will be protected. In this chapter, we shall explore how dispute resolution options can meet the obligations of both parties and facilitate the administration of justice in the Statutes and Regulations which govern civil aviation.

Building on the relationship between the State and the individual has been a major focus in this book, and the benefits and threats of management control by a prosecutor analysed critically, enabling us to assess the options for a solution, with a number of models which, together, may form a cluster of dispute resolution processes. The key issues of regulation by the UN Charter, giving rise to criminal proceedings commenced in domestic courts, must be analysed and compared with other forms of dispute resolution, not the least being mediation. Given the mischief of criminalisation which the Charter reserves to a domestic Court, we must be open-minded in an option, examining, and perhaps extending, the application of international law to criminal accountability, and with alternatives to prosecution in resolving disputes which otherwise may be regarded as unfair or intractable. While it may alarm the traditionalists in governments and powers, the solution of traditional litigation must be challenged, but challenged rationally, with its key feature depending on the benefits to accrue to Just Culture.

Most importantly, this chapter should not be regarded as a conclusion to the discussion of options, but as a start. In other words, let us set other hares running, and develop new options for the solution of the problem of criminalisation.

This work has concentrated heavily on cases in which offences have been dealt with through the criminal processes of domestic courts. It is right, therefore, that we should address this and ask, whether this offers a solution to the clash between criminalisation and a Just Culture.

DISPUTE RESOLUTION BY LITIGATION

The concept of justice is founded on the basis that the law is no more than a mechanism for solving particular problems that the parties have not been able to resolve by negotiation. But is justice all about fairness? Priority must be given to moral conduct

DOI: 10.1201/9781003489696-7

which must be observed by individuals towards each other in society. Violations of basic rights cannot be justified by arguing that such violations may produce economic or social advantages.

The practical problems have emerged simultaneously with the phenomenon of criminalisation, which has eroded the concept of fairness, but is seen, notwithstanding, as a trend in criminal accountability within the sovereign jurisdiction of domestic courts and underpinned by international law. Clearly, a new approach is needed to explore a solution; so we may start with the process of choice by sovereign States, in criminal litigation, and the underlying concept of justice which characterises it.

It is this which has challenged the global view of justice for the aviation professional. John Stuart Mill first published his seminal work *Utilitarianism* in 1861; his definition of Justice has not subsequently been improved:

> Justice is a name for certain moral classes of moral rules, which concern the essentials of human well-being more nearly, and are therefore of more absolute obligation, than any other rules for the guidance of life.[1]

This, therefore, is the single principle, founded upon the moral rules of society, against which we judge all actions. Within that very wide understanding, we have seen how society has evolved the notion of criminal accountability which has taken form in this book as the Mischief – clothed with all the negative characteristics which that noun implies, and which lies at the very root of the global aviation community's perception of criminalisation.

From our examination of the case studies, it is apparent that the aviation community's perception of a negative phenomenon of criminalisation is justified, in so far as it offends our understanding of criminal justice.

It is also apparent that this has been an emerging problem for some time; not long past, but, rather, the recent past. We have seen that the pilot in command has been criminally accountable for wilful or reckless acts which a Guilty Mind will attach to the circumstances of the offence and lead to criminal accountability, leaving the question of punishment to the Courts. This has been accepted as fair; but the phenomenon before us today has been formed and moulded by the normative ethics of contemporary societies globally, some of them having conflicting approaches, and conflicting priorities, resulting in convictions and sentences which trespass upon every concept of justice which is upheld by those peoples making up the aviation community who blamelessly choose to make this their career.

While the consistency of the regulation of criminal justice has been analysed exhaustively in contemporary literature, no solution has been found, or even attempted, to resolve the discord between the aviation professional and the State prosecutor, which has led to conclusions among professionals in the industry that the problem is intractable. A solution, therefore, necessarily compels the synthesis of a new approach.

The simple solution on a transnational basis would engage all States in a common process for criminal dispute resolution by the adoption of a single, common standard under new international law; the problem is that this must be adopted in domestic jurisdictions taking priority over their own laws. The attraction is that it provides consistency by taking further the philosophy identified in the concept of the Just

Culture that places the process of continuous learning and safety ahead of domestic prosecutions. The problems which taxed Eurocontrol in meeting the demands of sovereign criminal jurisdiction would present difficulties with this solution in promulgating new international law for the criminal prosecution of the pilot, for that very reason that it would trespass on the sovereign rights of States to refuse the primacy of international processes over their own legal systems, which they have the right to protect against external influence.

The promulgation of such new international law would, of course, be extremely expensive, time-consuming – and uncertain, as it would need States to agree to such a submission. While acknowledging the problem, a rational argument should be made for a uniform legal order, regulating criminal penalties for the fair accountability of aviation professionals for crimes which the process had determined their guilt. It is simply unacceptable to define an agreed set of principles for drawing the line between acceptable and unacceptable actions, without a consensus between the State and aviation professionals on the features defining the line and who gets to draw it.

The litigation process is the strict preserve of the Courts, who will manage a decision by interpreting the will of the law-makers according to the statutory provisions which govern the crime in question, either by an Act (under the English or Common Law process) or by the relevant Code (under the Roman process). In whatever regime they are found, however, they must submit to some basic rules of justice which underpin human rights.

Long before the European Convention on Human Rights, they evolved in the English legal system as the Rules of Natural Justice, largely uncodified which together define the basic assumptions of justice that all parties are entitled to expect in the legal process. They evolved largely through the control exercised by Superior Courts whose decisions must be followed in what are known as precedent cases and have therefore been applied in the English process through Common Law. Without statutory codification, they became Conventional Rules which responded to society's normative ethics of justice. There are, in fact, just two rules:

THE RULE AGAINST BIAS

The essence of a fair judicial decision is that it shall have been made by an impartial judge. *In R v Sussex Justices, Ex parte McCarthy*,[2] it was famously held that not only must Justice be done but it must also be seen to be done. The elements defining bias have been difficult to put into a principle, and the evidential issues proving this may be complex to establish; in deciding whether there has been bias, the House of Lords[3] endorsed the decision held by the Court of Appeal in *Re Medicaments*[4] that the test should address *whether the fair-minded observer, having considered the facts, would conclude that there was a real possibility that the tribunal was biased.* While the decision was applied in a civil case, the test translates well into criminal procedure.

This rule cannot be separated from the mischief of predetermining the case, an issue which has simmered away since the 1990s. In the 1996 case of *Kirkstall*,[5] Sedley J held that the principle applied to all litigation, in identifying apparent bias by reason of having an interest in a matter, demanding that a decision should be struck down if its outcome had been predetermined whether by the adoption of an inflexible policy or by effective surrender of their independent judgment. It is for this reason that a juror must be excused if they demonstrated a likely bias at the time when they are sworn in. Equally, of course, it must apply to a Judge who influences some predetermination in the course of the case.

THE RIGHT TO A FAIR HEARING

Each party should have the opportunity of knowing the case against them and of answering every allegation made against them. Both Prosecution and Defence must have the chance to present their version of the facts and to make submissions on the relevant rules of law, and each side must be allowed to challenge the evidence laid before the Court. This right to fairness also demands that neither side must communicate with the Judge behind the other's back; if there is an issue to address, then it is up to them to apply to the Judge in Chambers.

THE APPLICATION OF THE HUMAN RIGHTS ACT 1998

The 1998 Act merely serves to underline the validity of the English Rules of Natural Justice, as evidenced by Schedule 1:

> 1. In the determination of his civil rights and obligations or of any criminal charge against him, everyone is entitled to a fair and public hearing within a reasonable time by an independent and impartial tribunal established by law. Judgment shall be pronounced publicly but the press and public may be excluded from all or part of the trial in the interest of morals, public order or national security in a democratic society, where the interests of juveniles or the protection of the private life of the parties so require, or to the extent strictly necessary in the opinion of the court in special circumstances where publicity would prejudice the interests of justice.

We can apply these rules to the litigation process with a simple summary:

- The Judge must apply the Rules of Natural Justice in controlling managing the Trial process. In this, the Judge is no more than the voice of the legal system to which they, through their legal training and professional experience, are able to gain access.
- In the bulk of criminal cases, the Judge can discharge the Court's function simply by interpreting the Will of Parliament which passed the law in question; but they cannot influence it, because they must simply follow the existing rules and principles which the system has established.

While this nobly upholds the management of criminal justice, it fails to address the mischief that the law itself compromises the Defendant's right to fairness, as we have experienced in cases of criminalisation. As Judge Grenade concluded in the case against Captain Schröder:

> While I certainly do not discount the terrible consequences that have resulted from this negligence, what he has been convicted of is really a civil offense.[6]

BUT CAN WE HAVE CONFIDENCE IN THE PROCESS?

We might paraphrase the words of the Sentencing Council that confidence can be strengthened by improving public knowledge and understanding of a fair process, including among victims, witnesses and offenders, as well as the general public.[7] In this respect, we might consider the advantages of prosecutors with domain expertise in balancing fairness to meet the policy of Just Culture. Two examples will suffice.

The Netherlands may be justly proud as the cradle of Just Culture. In the Netherlands, the Public Prosecution Service (PPS) proclaims that Just Culture is taken into account in criminal justice.[8]

The PPS has a dedicated civil aviation arm, the Aviation Expertise Centre of the Public Prosecution Service (PPS) North Holland. Clothed with the advantage of domain expertise, this covers all Dutch criminal investigations relating to civil aviation with one national coordinating aviation public prosecutor, who is involved from the outset of a casualty and may order the aviation police to start a criminal investigation. They follow the same training programme as that of the Dutch Safety Board investigators, but, notably, they are separate from the investigation team, whose collection and analysis of date is central to the purpose of Just Culture. On its own website, it states:

> The PPS sees the aviation safety as an essential aspect of the general interest for our society. That aspect is therefore emphatically weighed in the settlement of aviation cases. By this, the PPS wishes to contribute to the advancement of 'Just Culture.'

It continues:

> In the aviation law and legislation a great weight is attached to this culture, which encourages the people concerned to report the incidents and to learn from it, without exempting them from normal responsibility.

That being said, the PPS is a prosecuting body, it is not a Judge, but the Prosecutor has discretionary powers to determine whether a prosecution should be pursued after the technical investigation has been concluded. Like the Code for Crown Prosecutors in the UK, they will consider whether a prosecution is in the public interest, in other words, whether it serves society as a whole. The issue of whether the Accused's human rights may be compromised do not feature in that; in particular, it will take into account their evidence gathered during the investigation process.

The PPS is meant to prepare the case management, therefore, to present to the Court where the individual's rights to a fair trial must be respected. How they define their powers of *exemption* and how they define *normal responsibility* as stated on their website are both unclear; it would appear that their discretionary powers are somewhat ill-defined for the aviation professional. However, the PPS does state *The PPS does not sentence every mistake*. This implies that the PPS has sentencing powers which overlap those of a Court in a prosecution, but a mistake may have been lost in translation.

In the UK, the statutory provisions defining the powers of the CAA are somewhat more clearly set out in Section 20 Cicil Aviation Act 1982:

> The CAA shall… have power to do anything which is calculated to facilitate, or is conducive or incidental to, the performance of any of its functions (including a function

> conferred on it by virtue of this paragraph)…. includes power to institute and carry on criminal proceedings.

The CAA therefore has the power to prosecute but must refer any judicial function to the Courts. There is a potential problem, in that the CAA is responsible for delegated laws from Parliament which it must also enforce, bringing it close to a conflict in the separation of powers, but just falling short of the final hurdle in which it is the Court that must try an individual and, if they see fit, convict them, thus in this case, the Executive is also the prosecuting body. But the author shares Professor Dekker's view that this offers a significant benefit. In the case of the CAA, the Prosecutor will have the domain experience of the expert, harnessed to the discretion of the Prosecutor to draw the line defining criminalisation, and together equips them to deliver confidence in making a decision based on an understanding of the aviation professional's split-second risk management and performance. The CAA may also demonstrate the advantage that the prosecuting authority will have end-to-end management of the safety information in a case, avoiding controversy between the technical and judicial authorities arising out of the protections of such information in Annex 13 of the Chicago Convention.

For all the author's reticence about combining the roles of Legislature and Prosecutor in one body (conflicting with the notion of the separation of powers), enforcement by the PPS (under the Roman system) and the CAA (under the English system) offers a solution in which the advantages outweigh the risks in the litigation system.

THE CRIMINAL LITIGATION PROCESS IN PRACTICE

Save ambiguities in understanding, once the decision has been made to prosecute, the individual will find themselves in the Court process, the management of which will be beyond the control of either Prosecution or Defence. It is this process which we must analyse, if we are to gather a clear understanding of the advantages and disadvantages of litigation.

In the adversarial system of English law, each party will present their case and protect their own interests in challenging the submissions and the evidence of their opponents. The issues of law involved will be argued and supported by the body of relevant evidence, that is, the evidence of non-expert witnesses on what actually took place as they saw or heard it, and the evidence of expert witnesses which assists the Court with informed opinion that enables it to reach a safe conclusion on the weight of evidence.

Each side is expected to present its own case and protect its own interests, subject to strict professional rules and duties to assist the Court in reaching fair and just conclusions, and any attempt to mislead the Court may be treated seriously, not only by the Judge but also by the lawyer's professional body, which is the Law Society for Solicitors and the Bar Council for Barristers.

In criminal trials in the Crown Court, the evidence is presented by the parties or their advocates to the Jury, consisting of 12 peers whose deliberations are jealously guarded secrets, who hear the evidence in chief of the witnesses, and their responses to cross-examination. Submissions in law are made to the Judge in the Jury's absence, for a consideration of the law is not within their remit. All they need know is the substantive law which makes a persuasive case on the evidence which they hear. In the lower criminal Court, less serious cases are heard before Magistrates without a Jury, but they must also apply the law to the weight of evidence before them, and an appeal can swiftly follow to the Crown Court if they get it wrong, when the case will be retried. The decisions of the higher appeal courts are based on the formal record of the Trial court – it does not typically consider new evidence, for appeals mostly are heard on a point of law.

In criminal cases, the Prosecution must persuade the Court that the case against the Defendant is proved Beyond Reasonable Doubt, as we mentioned in Chapter 1, by applying the evidence to the body of law. In practice this means that the Court must be absolutely sure – a phrase often employed by judges summing up to the Jury at the end of a criminal trial.

In England, Judges enjoy safeguards which ensure that they are entirely independent of the political power, so that they can discharge their judicial duties without fear of some personal consequence. They are appointed from the ranks of experienced lawyers, either Solicitors or Barristers, and selected for their good sound judgment based upon knowledge of the law, a willingness to study all sides of an argument with a clear degree of openness, an ability to reach a firm conclusion and to articulate the reasons for their conclusion, both as to conviction and as to sentence. In the latter case, they are assisted enormously by the Guidelines of the Sentencing Council, upon which they are wise to rely in great detail, for failure could lead to an appeal against sentence. But in any event, they are merely the controllers of the mechanism of the law; there is no room here for political influence.

In the United States, judges are typically attorneys who have been active in the political arena, and in fact have historically been elected by popular vote. This represents a major distinction between the American and English systems, but only as to their appointment. More importantly, this has been critically reviewed in the United States in recent years, and throughout all the States in the Union there is an acknowledged need to preserve judicial independence and need to deal with the Judge who cannot or will not properly discharge the functions of office. Appointment to the Supreme Court is derived from more political grounds, however influenced by the President, before the Senate Judiciary Committee then takes a vote.

Under both systems, Judges have constitutionally held one role in life when applying Statute Law: they can only interpret the Will of the Legislature. In practice, this means identifying the effect that Parliament intended when passing the Act in question. Judges have no power to make a finding as to whether Parliament was right or wrong in its intention; this remains the case today in the UK, and is even enshrined in the Human Rights Act, which protects the Judiciary from exposure in cases where the Law of Parliament must be interpreted, even at the cost of an individual's personal rights.

There is no doubt, that judges in both systems are likely to have areas of professional expertise, but there is no overarching authority which demands that they preside over a case in which they have domain expertise. Judges can make mistakes, and if they make an error in law (or applying the facts wrongly to the law) they can be appealed.

However brief, this conveniently sums up the concept under the adversarial system, which differs from the inquisitorial system practised in jurisdictions which have followed the traditions of Roman law, in which a Judge (or Examining Magistrate) investigates actively the case in front of them, in order to determine the case facts partially or fully depending on the seriousness of the offence. Unlike the process in the adversarial system, though, the investigation authority is expected to collaborate and co-operate with the judicial authority in order to establish the facts based on their expertise, in respect of which Mildred Trögeler concluded that it would be beneficial to educate the judicial authority in the domain expertise of civil aviation in order to gain a better understanding of the operational and technical circumstances. Such a solution may be feasible for Examining Magistrates under the Roman system, before they analyse the weight of expert evidence from both sides. She makes the point that integrating domain expertise in the role of the Judge would increase credibility and the justice of the subsequent prosecution and, in the Roman inquisitorial system, she may well have a point. The advantage in the adversarial system under English law would be more obscure, in which Judges are there to manage the procedure that arises between Prosecution and Defence.

The Roman system is designed to determine the case management and procedure in which the Trial is to be conducted, and as such will recommend a Trial only if the Court is sure that there is sufficient evidence of Guilt. The entire body of evidence must be made available to the Defence in pretrial proceedings, who will have the powers under the European Convention on Human Rights to be heard in a fair trial, albeit without a Jury of their peers.

In practice, criminal litigation suffers from a number of characteristics that discourage its employment in the criminalisation of offenders:

- The outcome is entirely out of the control of the parties – it must be, for the Rules of Natural Justice to prevail, but the loss of control also means the loss of any power to reach a solution that meets the priorities of the Defendant.
- Associated with this, is the risk that no case can be considered 100 per cent watertight by any party.
- The process is expensive, demanding the use of Counsel and, inevitably in aviation cases, expert witnesses.
- As the case studies demonstrate, the outcome may follow the laws and procedures in the sovereign jurisdiction of the Court, but it may not follow the expectations in which the aviation professional must trust; it may not be justice, but it is the law.
- The process takes a long time to reach judgment.
- The downstream consequence on the perception of a just outcome may have negative effects on the element of trust upon which the aviation professional is entitled to rely.

THE CAA'S NEW APPROACH TO PROSECUTION

In the Civil Aviation Act 2012, Parliament laid the foundations for a power for the CAA to impose civil sanctions, as penalties that can be imposed without going to Court, as an alternative to taking action in the criminal courts or to warning letters. This followed a consultation process that accepted the unsatisfactory current enforcement methods of criminal prosecutions and suspending or revoking a pilot's licence at one end of the scale, and unenforceable warning letters at the other, but with nothing in between, suggesting an alternative in that *The Civil Aviation Act 2012 amended the Regulatory Enforcement and Sanctions Act 2008 to enable an order to be made giving the CAA access to a range of civil sanctions to use in addition to their existing penalties.*[9] The CAA highlighted current disadvantages that court action can be slow, cumbersome and expensive, and a criminal sanction may be a disproportionate way of enforcing certain aviation related offences. Civil sanctions were raised as a more flexible and proportionate alternative.

In their consultation with the wider community, the Executive sought evidence, views and comments from those who may be affected by the introduction of civil sanctions for use by the CAA as an alternative to existing enforcement measures, which currently range from warning letters to criminal prosecution.

The CAA's policy objective envisaged new regulations giving it a range of powers to impose civil sanctions, which would *allow the CAA to address the lack of proportionate and cost-effective enforcement action by using the new powers for appropriate enforcement.*

Its long-term objectives envisaged *higher standards of aviation safety, by helping to bring offenders back into compliance sooner, discouraging non-offenders from offending at all, and enabling to the CAA to act swiftly and flexibly to prevent breaches in regulation that may lead to serious harm.*

This certainly would meet certain concerns held by aviation professionals that litigation would be extended by an alternative route from criminalisation to civil sanctions – provided, of course, that the human rights of the offender would be respected with the right to a fair hearing.

As long ago as 2005, the Hampton Review[10] of the regulatory system concluded a wide-ranging analysis of the administrative burden of regulation in terms of the cost in time or money of regulators' inspection and enforcement activities. It made a number of recommendations, for all regulatory activity that, among other things, should be open to scrutiny, balanced by including past performance and potential future risk, using all available good quality data, and that it should be implemented uniformly and impartially. It was followed in 2006 by Professor Richard Macrory's review of Regulatory Justice in the light of the Hampton recommendations,[11] underlining the point that regulators often had to choose between issuing a warning letter or caution at one end of the scale, and taking criminal proceedings at the other, with nothing in between. Macrory pertinently also observed:

> In instances where there has been no intent or wilfulness relating to regulatory non-compliance, a criminal prosecution may be a disproportionate response. However, regulators may not have any other sanctions available within their toolkits. If the actual or potential consequences of the regulatory non-compliance are serious, regulators may want to take some action, and the public may expect the regulator to take some enforcement action, but the only option available is a criminal prosecution.

Macrory particularly addressed the unsatisfactory application of regulatory offences giving rise to strict liability, dispensing with the need to establish a mental element involving intention, recklessness, *or even negligence on the part of the offender.* He sought the advice of Professor Simester of Nottingham University who concluded that the majority of core offences surveyed impose strict liability:

> Serious regulatory breaches (such as those giving rise to significant damage or significant economic gain from non-compliance) which are strict liability offences, but where there is no evidence of a deliberate or reckless flouting of the law or gross negligence by the business concerned would, in my view, be more appropriately dealt with by the imposition of an administrative penalty.

That being said, Macrory underlined the need to retain the criminal prosecution process for serious breaches which relied on evidence of intention or recklessness, where there was evidence of a criminal mind, as well as reposing discretion on regulators to pursue a prosecution in cases where they find evidence of gross negligence and where the actual or potential consequences of the breach are so that the public interest test in the Code for Crown Prosecutors demands a criminal prosecution.

Perhaps the most pertinent of all responses to the Macrory review came from the UK's Justices' Clerks' Society, a collective body committed to improving the quality of justice in Magistrates' Courts; at the sharp end of the criminal prosecution process, their opinion is particularly valuable:

> A regime of administratively imposed sanctions is particularly suited to defaults that are highly technical or where the resultant offences are ones of strict liability. Courts on the other hand are best suited to making judgements on the basis of fault, particularly where intention or recklessness are necessary ingredients.[12]

A considerable amount of work has been conducted into which offences would be suitable for the penalty of a civil sanction. Of 641 regulations identified, many apply to low-level administrative breaches, defined in the Air Navigation Order 2016, section 265 (1) stating the current position on enforcement that they would be subject to criminal punishment in the Court process:

> If any provision of this Order, any regulations made under this Order or a Safety Regulation is contravened in relation to an aircraft, the operator of that aircraft and the pilot in command... is (without prejudice to the liability of any other person for that contravention) deemed for the purposes of the following provisions of this article to have contravened that provision.

At least the due diligence defence is still available, as defined in subsection 2:

> A person will not be deemed to have contravened a provision specified in paragraph (1) if the person proves that the contravention occurred without that person's consent or connivance and that that person exercised all due diligence to prevent the contravention.

Such violations meriting a civil sanction would include cases such as:

- Failing to return a certificate of registration to the CAA.
- Failing to inform the CAA in writing forthwith of any change in particulars, destruction of the aircraft or termination of a demise charter.
- Holding oneself out as a person who may offer flights in a UK-registered aircraft for the purpose of public transport or commercial air transport without holding an air operator's certificate.

The more serious cases which are deemed not suitable for civil sanctions include cases such as:

- Recklessly or negligently acting in a manner likely to endanger an aircraft or any person therein.
- Reckless or negligently causing or permitting an aircraft to endanger any person or property.
- Permitting a person to act as an air traffic controller without such person holding an appropriate licence and being satisfied that the person is competent to perform his duties.

By moving away from the process of criminalising violations, the associated mischiefs of cost and administrative burden identified by the government are mitigated, while the underlying benefits of a Just Culture policy are enhanced, however remote that may have been in the government's intentions.

WHAT OTHER OPTIONS PRESENT THEMSELVES?

Of course, we are ignoring the obvious, that the adversarial and inquisitorial systems suffer more from what divides them than what unites them. The resultant inconsistencies in the prosecution of aviation professionals therefore will continue to compromise the trust in the litigation process from an international point of view. Michaelides-Mateou and Mateou[13] propose a solution with alternatives in an Aviation Arbitration Tribunal, or else a European Aviation Court. As regards the latter, the uncertainties in creating a model for an international Court are challenging:

1. Any such option will require a treaty which would have to adopt a process of prosecution which is acceptable to parties which practise the adversarial and the inquisitorial systems in any combination. This is heavily mined by hazards, not the least of which involve the conflict between the two systems in how evidence is managed and presented.
2. The extortionate cost and time expended in getting a draft instrument to the stage of a ratified Convention might be seen as an incentive for revisiting other options.
3. Once the treaty is in place, the management of the Court would have to be paid for, and parties would have to contribute judges with the expertise in which both the State of Occurrence and the State of Registration had confidence.

However, the solution of an Arbitration process may offer a more positive option that can be applied both in domestic and international cases.

A CRIMINAL ARBITRATION PROCESS?

An alternative dispute resolution (ADR) process has long been observed in arbitration. Arbitration is generally regarded as a process exclusively reserved for civil law. Exceptional situations arise when an arbitrator is confronted with evidence of crimes, such as bribery or money laundering, while deliberately giving false testimony may give rise to a referral for criminal investigation; but the procedure is blurred.

The whole structure of international regulation relies on consensus: not just in ratifying terms but also in resolving problems. For this reason, the United Nations sets a lot of store by arbitration as an alternative to the Court process. Indeed, if the parties cannot agree on a forum, arbitration is obligatory; so, for example, the Chicago Convention empowers the Council of the International Civil Aviation Organization (ICAO) to cooperate together and share their skies to their mutual benefit, but Chapter XVIII gives authority to the ICAO Council to decide upon any disagreement between Contracting States regarding the interpretation or application of this Convention, if such a disagreement cannot first be settled by negotiation. In this respect, it is an arbitration process, pure and simple.

Clearly, the process is therefore applied for disputes between States in the case of civil aviation, but how could it work in the domestic system of criminal accountability?

Arbitration may be defined broadly as a procedure whereby parties in a dispute refer the issue to a third party for resolution, rather than take the case to Trial in the Court Litigation process. In domestic regimes, it has been seen as a private process, the authority for which is derived by the agreement of the parties, although the arbitrator will be obliged to comply with the Rules of Natural Justice.

The Arbitration Act 1996 repealed Part 1 of the Arbitration Act 1950 and the whole of the Acts of 1975 and 1979, and is intended to provide a framework that encourages the use of Arbitration in preference to litigation, following the Model Arbitration Law adopted in 1985 by the United Nations Commission on International Trade Law. The provisions of this Part are founded on the following principles, and shall be construed accordingly:

- the object of arbitration is to obtain the fair resolution of disputes by an impartial tribunal without unnecessary delay or expense;
- the parties should be free to agree how their disputes are resolved, subject only to such safeguards as are necessary in the public interest;
- in matters governed by this Part, the court should not intervene except as provided by this Part.

Unlike the Court process, in which the parties have no right to pick and choose their Judge, who may or may not have any domain expertise, the arbitrator certainly will be appointed for professional expertise in aviation and can be a lawyer with much experience, harnessing technical knowledge to legal expertise. They therefore can be confident in making decisions which are safe.

Arbitration has significant benefits, in that it is a private process, and with an expert as arbitrator is more likely to deliver trust, meeting the worries of the individual or corporate body who may suffer from consequent publicity. The disadvantage in the criminal process lies with enforcement; it does not have powers of jurisdiction and sentencing about how offenders are dealt with before and after conviction under the Criminal Justice and Courts Act 2015 and is apparently outside the Criminal Procedure Rules 2020 which regulate criminal court procedure in magistrates' courts, the Crown Court, the Court of Appeal and, in extradition appeal cases, the High Court.

There are good reasons why the parties would wish to arbitrate in a legal dispute:

- Arbitration remains a cheaper ADR process to litigation, insofar as the parties incur lower bills of costs, and the process takes a shorter time to resolve than in a protracted case in law. This arises partly because of the greater informality in arbitration as regards the management of evidence, and explaining the technicalities, for the chosen arbitrator will have domain expertise. For this reason, the party's ability to choose the right person for the job is an important advantage whereas they had no influence whatsoever in the appointment of a Judge.
- In general, the powers of arbitrators fall short of those conferred upon Judges in Litigation, as a result of which the prosecuting authority may be discouraged by a cost-benefit analysis, while the defending individual or company may be encouraged that will not be caned as hard as they might be as the result of a judgment. That being said, once the Arbitrator has been appointed, the decision-making process is as much out of the parties' control as it is in Litigation. An appeal from an arbitrator's decision must be available to the parties but only on a point of law. It would be almost impossible to challenge an arbitrator's decision on a point of fact.
- Of considerable importance is the confidentiality of arbitration proceedings. Allegations of bad faith, misconduct, negligence and issues damaging to a party's reputation frequently arise in any dispute, and the forum of arbitration will hear such evidence in privacy. It is an essential characteristic that such evidence will be heard in a Trial in open court, which means that the press and public will be allowed to hear the evidence and cross-examination, and as a Court of Record will have access to the issues and to the judgment. In the arbitration tribunal, the evidence is heard in private. Where the parties agree, however, the arbitrator will permit the award to be published.
- The more informal approach is less damaging to future relations than the endurance test of litigation, encouraging a concept of restorative justice, bringing offenders closer to the communication process leading to an understanding of their wrong-doing and finding a positive way forward.

The CAA actually has background knowledge of ADR, in that they act effectively as arbitrators themselves, as part of the UK's commitment to act to increase consumers' access to ADR, meeting the demand of resolving consumer complaints giving rise to claims in commercial or civil law in the provision of aviation services. It is noteworthy, however, that this is limited to consumer disputes.

Even under Current Law, arbitration tribunals must take account of criminal laws and procedures, so the criminal domain is not unknown to them; but they need not strive to meet them to the letter. The arbitrator is obliged, rather, to resolve the dispute in accordance with their remit, under which they were appointed, enabling them to reach their objective of an enforceable award by their own skill and knowledge, resulting in a decision which may carry more trust than had it been tried in Court, and for those offences identified as being suitable, which accords with the underling benefits of a Just Culture.

It would take a bold decision to rationalise the idea of criminal arbitration into a proposal which a government could accept. Given the CAA's powers to apply civil sanctions to certain criminal cases, however, it takes one further step to transfer the determination process to an arbitration tribunal. Once again, we meet the dividing line between a case suitable for a civil sanction and a criminal case which demands proof beyond reasonable doubt of the Defendant's *mens rea* and *actus reus*.

CRIMINALISATION: AN INTERNATIONAL PROBLEM NEEDS AN INTERNATIONAL SOLUTION

We have explored the options for criminal dispute resolution in domestic regimes but, in fairness, there are significant opportunities for the resolution of criminal cases with an international element in civil aviation, based on the rights and obligations of the State of registration and the State of occurrence. The international outrage following the Italian criminal trial of Captain Chafik Al Gharbi of Tuninter Flight 1153 highlighted the frustration of the aviation world with the unassailable position of State Sovereignty in criminalisation, but international law, itself, offers a solution if only the States Parties can place trust in their Conventions to deliver something that they can live with.

It is important in this work to offer a solution which restores an ethical application of the criminal law to the aviation professional on a transnational basis. In this sense, the ethics underpinning the position of the pilot in command must be applied broadly and it is worth repeating that the criminal law must be applied subjectively to the pilot, raising the need to establish a guilty mind, but from which the process has drifted away in the evolution of the crime of criminal negligence. When the society of the State of occurrence reflects on the ethics which determine the pilot's accountability, it may not be limited to judging whether the pilot's acts are morally right or wrong. Nardin and Mapel[14] distinguish some ethical judgments, as being moral in the restricted sense that they involve the application of principles of what society holds to be proper or improper conduct, surely to be determined by a subjective test upon the pilot's *mens rea*, by comparison with judgments which are concerned with acts which are permitted or condemned according to the features which characterise those acts, in other words, the circumstances alone which may not demand a guilty mind. Such judgments in the criminal law may vary dramatically according to the values and interests of the State of occurrence and suffer intractable conflict with those of the State of registration which clothed the pilot in command with their authority in the first place. In this context, the trend in recent years, of requiring mere strict liability for offences carrying increasingly severe penalties, touches a sensitive nerve for the criminalisation of the aviation professional.

In Chapter 3, we discussed the compelling argument for a solution that implements globally a common ethical system based on the individual's human rights. A common ethical system across States necessarily limits a State's inalienable freedom of choice, but there is nothing to stop a State from exercising the equal freedom of picking and choosing which legal principles and processes it admires in other jurisdictions; and in negotiating an international Convention, they can make such choices before adopting it. The risk, of course, is that they all pick and choose a common definition of criminalisation that perpetuates the mischief. But at least we have an international response with a common process for accountability of individuals that can be applied in domestic courts, and its logical progression leads to a solution in which the accountability can be resolved between States on a Convention basis by an arbitration process that transcends sovereign borders.

CAN WE RELY ON INTERNATIONAL OBLIGATIONS BETWEEN STATES?

This option necessarily arises in the circumstance, either that the State of registration of the aircraft has broken its obligation to the State of occurrence, or vice versa, for example, by interfering in the lawful flight of an aircraft, or by prosecuting the pilot in command under its domestic law which did not apply on a jurisdictional basis or in a strict liability case that trespasses on criminalisation, or by some other act by which the innocent State's rights had been infringed according to the obligations under the relevant Treaty.

Article 33 of the United Nations Charter gives pre-eminence to the principle that disputes must be settled peacefully, leaving the choice of means to the parties. (If the parties cannot agree on a forum, arbitration is obligatory.) If the dispute cannot be arbitrated effectively as envisaged by the Chicago Convention, then it can still be referred for judicial resolution by the principal judicial organ of the United Nations, the International Court of Justice (ICJ), whose powers are defined in Chapter XIV of the United Nations charter. Although the charter does not employ the specific words, it sees itself as a 'world court.' The ICJ has a dual jurisdiction, the relevant one to this work being to decide, in accordance with international law, disputes of a legal nature that are submitted to it by States – which the ICJ defines as a disagreement on a question of law or fact, a conflict, a clash of legal views or of interests.

As a general rule the ICJ does not have jurisdiction to hear disputes unless both States Parties have agreed to submit to its jurisdiction on the matter. The ICJ does not bask in the warmth of universal acclaim, however. In its dispute resolution function, it does not permit access by bodies other than States who have signed up to the UN Charter. This may be inconsistent with the human rights of the pilot in command or the air traffic controller who should naturally be entitled to join as a party, and therefore be heard, in proceedings in which they will be accountable for their conduct. A compromise solution enables the UN General Assembly and the Security Council to seek advisory opinions from the ICJ on any question of international law, while other non-States Parties may do so in respect of matters falling within their specific competence. For the representation of the pilot in command or the air traffic controller, though, this solution is unavailable; in any event, the likelihood that the awesome majesty of the Security Council would be troubled to gaze upon some trifling maritime dispute is not a realistic contingency.

A highly relevant issue in this context involves the Convention on Offences and Certain Other Acts Committed on Board Aircraft, the Tokyo Convention 1963. The Convention establishes a uniform approach to acts on board aircraft which are offences against penal law, or which may or do jeopardise the safety of aircraft and persons or property on board, or good order and discipline on board. The Convention includes provisions relating to the exercise of jurisdiction and detailed rules as to the powers of the pilot in command.

Articles 3 and 4 of the Convention ensure that the State of registration of an aircraft has jurisdiction over offences committed on board, but it also acknowledges the possibility that other States may claim jurisdiction over such conduct, and that there may exist overlapping or concurrent jurisdictions. It does not exclude any criminal jurisdiction exercised in accordance with domestic law, and Article 4 permits a Party, although not the State of registration of the aircraft, to interfere with an aircraft in flight in order to exercise criminal jurisdiction certain cases:

- The offence has effect on its territory.
- The offence has been committed by or against one of its nationals or permanent residents.
- The offence is against its security.
- The offence consists of a breach of any of its rules or regulations relating to the flight or manoeuvre of aircraft.
- The exercise of jurisdiction is necessary to ensure its observance of any obligation under a multilateral treaty.

Article 24 offers a dispute resolution process in that *any dispute between two or more Contracting States concerning the interpretation or application of this Convention which cannot be settled through negotiation, shall, at the request of one of them, be submitted to arbitration. If within six months from the date of the request for arbitration the Parties are unable to agree on the organization of the arbitration, anyone of those Parties may refer the dispute to the International Court of Justice by request in conformity with the Statute of the Court.*

ARBITRATION: WHAT CAN POSSIBLY GO WRONG?

The arbitration of disputes between States has a long history, sadly not always a happy one, with a discouraging effect on the element of trust in the process. Perhaps its most notorious application was in the wake of the American Civil War which was fought between 1861 and 1865. With the end of hostilities, a victorious Washington demanded redress from Britain for the unforgivable sin of giving tacit support to the wrong side, claiming damages for losses inflicted by the highly successful armed merchant cruiser CSS *Alabama*, which had been built in Britain, as well as a number of other ships supplied to the Confederacy. Britain would go so far in the interests of diplomacy, but would not accept liability for the incredible quantum of damages claimed by Washington, far outweighing reasonable compensation and amounting to more than the British National Debt, which would actually have given them a profit for waging the war. But the United States was in no mood to negotiate on its claim, and the only solution was litigation. The trouble with a dispute resolution process was that there was no precedent in any treaty upon which they could call, so one had to be agreed. The Geneva Conference followed as the treaty establishing an arbitration process to resolve the dispute.

The Arbitrators found on a majority of 3 to 2 that there was no ground for awarding such sums to the United States. They also claimed for future losses: that is, the loss of profits which Northern vessels would have earned in their trading careers, but for the intervention of the Confederate raiders. The Tribunal was unanimous: there was no ground for such an indemnity. Refusing to allow another evil, the Tribunal set aside the many double claims which had been made for the same losses, as well as inflated sums for *gross freight* in so far as they exceeded the acceptable net freight. They did, however, allow interest at a reasonable rate on all the damages that had been awarded.

The question of assessing quantum, or the amount of the damages, had been delegated in the Treaty to a Board of Assessors, but the Arbitrators had the power to award a gross sum instead of referring the proceedings on further. They duly decided on a gross sum of *fifteen millions five hundred thousand dollars in gold as the indemnity to be paid by Great Britain to the United States for the satisfaction of all the claims referred to the consideration of the Tribunal.*

It was an incredible sum, which, in fact, was so much in excess of the claims actually made by the United States that Washington was faced, literally, with an embarrassment of riches and, in December 1872, the Senate passed a Bill to create a commission to adjust the compensation due to claimants that would be discharged out of the indemnity fund awarded by the Conference.

It was, of course, a shocking example of unfairness; but the arbitration process had been established by agreed terms, so, notwithstanding the findings in law dismissing many of the claims, there was no basis for an appeal on the quantum of the gross sum awarded by the Arbitrators. The British had to keep good faith with the process, like it or not, and in that nothing has changed since then.

THE INTERNATIONAL PROCESS TODAY

Article 24 of the UN Charter compels States to refer an unresolved dispute to arbitration before recourse to the ICJ, which has a reputation for suffering two major ills, of time and cost. Between 22 May 1947 and 29 April 2024, only 195 cases were included in the General List, which conveys something of the huge complexity in being a party to proceedings in the ICJ, and the associated costs can really only be borne by States Parties who can afford them – or justify them. Moreover, a delinquent State merely has to deny jurisdictional status to the International Tribunal, in whatever form it takes, and the process becomes farcical, as Russia has proved in maritime cases such as that of the *Arctic Sunrise*. In that case, the Netherlands as Flag State claimed that the actions of Russia, in boarding, arresting and detaining the vessel constituted a breach of obligations owed to the Netherlands as Flag State regarding freedom of navigation and the right to exercise jurisdiction over the *Arctic Sunrise*. Russia promptly announced that it did not accept the arbitration proceedings initiated by the Netherlands, and that it did not intend to participate in the request for provisional measures. Russia simply failed to enter an appearance and defend the case, but the International Tribunal[15] held that the Netherlands should not be put at a disadvantage because of the non-appearance of the Russian Federation in the proceedings; hence, Russia was a party to the proceedings with ensuing rights and obligations and the Tribunal was not precluded from proceeding with the case against it. While the parties were obliged to seek negotiation before such avenues were exhausted, the Tribunal concluded that the Netherlands had tried, but could not go any further.

Eventually, in 2017, Russia was ordered to pay damages of €5.4 million to the Netherlands for the capture and detention of the *Arctic Sunrise*. Despite Russia's vehement rejection of the Arbitration and its outcome, the parties continued to negotiate and, in 2019, five years after the seizure of the vessel, the parties issued a joint statement in which Russia continued to deny liability but agreed to pay €2.7 million to the Netherlands in settlement of their claim.

We can draw a swift conclusion to these questions: States Parties in international arbitration must be bound by their agreement, but a State lacking in good faith, or, indeed, good will, puts into severe doubt whether we can rely on international obligations between the States.

IS THERE ANOTHER FORM OF ALTERNATIVE DISPUTE RESOLUTION?

Pure forms of ADR require consensus to apply to the form and outcome of the process – that is, the whole experience is controlled by the parties themselves, from the scope of the dispute to be addressed, to the final settlement. While alternative dispute resolution takes a number of forms, the flagship is Mediation.

Mediation is fairly new to the litigation process in many parts of the world. Its birth was a matter of necessity rather than choice, as the cost of litigation in terms of money, time and uncertainty escalates out of acceptable terms to the parties, leading them to consider a new approach. Its application in the twenty first century world has evolved from experience in the United States in the last quarter of the preceding century, where the legal process had led the world in the arena of expensive litigation. But mediation has met with mixed results in different regimes. Despite its considerable success in the United States, it has not been so widely embraced in the UK. At first sight, one might wonder why. Litigation costs have an endless capacity to escalate out of control. Nowhere has this been more keenly felt, than in the United States in recent years, where they tackled the problem head-on, and found the solution: let the parties keep control of the situation and explore a settlement – not with a judge who will rule between them, but with a trained neutral person who will guide them. Mediation has been a feature of litigation in the United States for some 50 years now – a feature which has been found time and time again to be a cost-effective alternative to litigation. Even in those apparently intractable cases which have gone to judgment and subsequently been appealed, current statistics show that a large percentage of those are then successfully mediated before the appeal hearing.

Mediation's success story in the United States has been so compelling that it has been embraced by the British government as a vital tool in modernising the Courts system in this country, and procedural rules have been in place since April 1999 actively encouraging the use of Mediation and ADR in every civil case. It is a fact that in the year to May 2000, the number of cases in England which were mediated as an alternative to going to Trial increased by 140 per cent. But its popularity waned.

The application of Mediation in the international context, readily embraces the idea of dispute resolution without boundaries: the parties need not be States which have signed up to any key treaties, and there are no rules of procedure limiting just how flexible a solution may be reached, in order to meet the parties' priorities, not just resolving the event which has taken place but also establishing some future relationship which gives the parties confidence in each other. The key is in the parties' own power, to reach some guided settlement by negotiation.

It follows that the starting point in the dispute resolution process is a simple discussion between the parties themselves to explore a settlement that everybody can live with; after all, a problem becomes a crisis when negotiations break down and the critical function of communication is lost. So the alternative must be available at the earliest possible opportunity, and it must remain available until its possibility of success has been exhausted – that is, up to Trial – to produce a result that ends up with a commitment to a solution.

Hence, the first treatment option in dispute management can be defined as Negotiation. Although an informal process without a rigid structure, there are very few critical issues, but they are indispensable:

- It is essential to start by separating personalities from the problem; getting personal merely leads to entrenched positions.
- Now the negotiation must focus on interests, not positions. In other words, the discussion must address two questions, which demand understanding – and respect – from both sides:
  i. What are their priorities?
  ii. What are our priorities?
- From here, it is simpler to identify the areas of agreement between the parties, and clarify the issues that divide them.
- This leads to the step of finding the common interest, leading to a solution that everybody can live with.

The key factor is communication; but when communication is lost, the negotiation function is lost.

Mediation is nothing more than an extension to the process of negotiation. Once communication has broken down, and the people are unable to reach a solution to a problem themselves, it makes sense to seek the assistance of a neutral person – the Mediator – who is especially trained and skilled in the process. By working with the parties, the Mediator helps them to clarify the issues that divide them and identify the matters on which they agree, which makes it easier to explore a solution, which will not be ideal for one party or another, but will be something with which all of them can live. In this way, the parties remain in control of the negotiation.

The Mediator, understandably, must guide the process by relying on their own expertise in the issues, and by exercising to the full their energy, patience and persistence. That really is all they have, for there are no rules in Mediation – what the parties want, they get. The Mediator's function is to facilitate the negotiation, by establishing the foundation of confidence. Without any issues of reliance upon precedence or legal regulation, confidence is essential for the Mediation to work, because the parties must have confidence in the process, and confidence in the Mediator to steer the process – and the parties – to a solution to their dispute.

At the core of the parties' confidence is their respect for the Mediator's neutrality: the Mediator has no axe to grind one way or the other, and so the parties can rely on the vital commodity of confidentiality when they are discussing their position with the Mediator behind closed doors. This aspect of confidentiality is crucial to the process, because the Mediator must know what the parties' priorities are – what is important to them – if there is to be any hope of finding a solution. Very often they have kept very quiet about these priorities, which would otherwise betray some weakness in their case or their bargaining position, and they would not be at all thrilled if these were disclosed to the other side; but the Mediator will know what their needs are, and will skilfully keep this in mind without disclosing any confidences unless the party in question authorises him to do so.

The pressures that build up in a dispute frequently force cases out of the hands of the parties, even if they did not want it to escalate; this can be particularly dangerous to their priorities when they have, want, or may need to have ongoing relationships of some sort, such as that between States. Time is invariably an important issue, for they have a mutual interest in a relatively swift resolution, while litigation appears long, expensive, and risky for both, and the outcome is out of their control: experienced lawyers are aware that no case is really more than 80 per cent certain, so much can go wrong in the end. But both parties feel that they are being pushed into litigation by a whirlwind of accusations and arguments. Driving it all, of course, is the clash of the normative ethics of their respective societies, generating strong emotions and venting negative feelings which prevent them from looking objectively at ways to meet underlying interests and jointly solve problems.

The resultant effect of these pressures forces a breakdown in communication between the parties, and lost communication is the mid-wife of Litigation. By comparison, the mediator will help them to communicate, and guide them to recognise their own and each other's legitimate priorities; from this foundation, they can clarify those areas of mutual agreement, and isolate the dangerous areas of disagreement, leading them to design options which reconcile and meet those needs, creating a solution which would be beyond the power of the court to impose, but which can be closed in the form of a binding agreement in which the parties can have confidence in the future.

Experience in recent years of high-level political international mediation and the growth of international trade – and the damage caused when it is interrupted have opened opportunities for mediation, even in what have been viewed as intractable conflicts. Jacob Bercovitch[16] characterised intractable conflicts as being driven by protagonists who have a strong sense of identity; by application in context, this associates the phrase with the classic scenario of international flights in which sovereign states harbour a grievance against each other, be it economic, political or, indeed, otherwise sensitive, such as the damage to one party's environment or the human rights of the other party's representative, the pilot in command. The successful mediation of intractable conflicts in the international arena, in which the antagonists reach a jointly acceptable outcome without assuring their political suicide with their electorate, raises the real option that a dispute arising out of an alleged claim by the state of occurrence against the state of registration, can be mediated without the cost of litigation and without the necessity for any new international law whatever.

The question remains, however, as to how realistic the notion presents itself, in the context of this work. Realistic minds have expressed the opinion that the conflicts between the State of Occurrence and the State of Registration are unlikely even to allow them to agree on a mediation process, let alone to pursue it, when the prospect of a final solution is entirely dependent upon the parties' will to find one, and to keep to its promises – for mediation has no teeth.

But that puts mediation in the context of being an alternative to an international tribunal. Mediation may be observed to offer an additional option, as a stage in the process following the breakdown of negotiations but before litigation before a tribunal. After all, it is a consensual, non-legal remedy, so if the parties cannot reach agreement, they can proceed to Trial. The UK government has embraced this as a key part of the management of civil cases, and itself has published a commitment, on its own behalf, underlining a key objective to *actively consider and attempt the use of alternative dispute resolution techniques whilst making going to court to resolve the dispute a last resort wherever possible.*[17]

In the light of this, mediation could be built into the option for a solution, as a means of facilitating some negotiated process, which, if unresolved, would enable the parties to proceed to the next step, of testing their case in Court.

For the pilot in command, however, this option only presents the opportunity of expediting a possible resolution of the dispute between the State of Occurrence and the State of Registration, regarding their international commitments to each other. If the States cannot reach a resolution under the existing regime, then these options remain unresolved as solutions for the pilot in command, whose criminalisation will continue.

NOTES

1 Mill JS, 1863, *Utilitarianism*, Parker, Son & Bourn, London, p. 87.
2 *R v Sussex Justices, Ex parte McCarthy* [1924] 1 KB 256; [1923] All ER Rep 233.
3 *Magill v Porter* [2002] 2 AC 357.
4 *Court of Appeal in Re Medicaments (No. 2)* [2001] 1 WLR 700.
5 *R v Secretary of State, ex parte Kirkstall Valley* [1996] 3 ALL ER 304.
6 *United States of America v Wolfgang Schröder* [2007] United States District Court, Alabama Southern District (currently unreported).
7 See www.sentencingcouncil.org.uk.
8 See www.prosecutionservice.nl.
9 CAA, 2011, Civil Actions Impact Assessment.
10 Hampton P, 2005, *Reducing Administrative Burdens: Effective Inspection and Enforcement*, HM Treasury, London.
11 Macrory, 2006, *Regulatory Justice: Sanctioning in a Post-Hampton World Consultation Document*, Better Regulation Executive, London.
12 Response from Justices' Clerks' Society to the Macrory Review, February 2006.
13 Michaelides-Mateou and Mateou, supra.

14 Nardin T and D Mapel (eds), 1992, *Traditions of International Ethics*, Cambridge University Press, Cambridge, p. 326.

15 The 'Arctic Sunrise' Case (*Kingdom of the Netherlands v Russian Federation*), ITLOS Case No. 22, Order of 22 November 2013.

16 Bercovitch J, 2004, International Mediation and Intractable Conflict, Beyond Intractability, in G Burgess and H Burgess (ed), *Conflict Research Consortium*, University of Colorado, Boulder.

17 Attorney-General's Office, 2011, *The Dispute Resolution Commitment*, Ministry of Justice, London.

8 Help! Getting Out of Jail
Some Practical Advice

OVERVIEW

Having spent a hundred thousand words on examining what can go wrong, we might conclude with a few words of advice on what to do when it does go wrong.

THE FIRST GOLDEN RULE

There are just two golden rules. The first is: Get professional indemnity insurance. NOW. The Internet contains numerous sites of insurance providers for aviation professionals. This book plays no part in making recommendations, but whichever provider you choose, your cover will only start after the insurance contract has been closed and the premium paid; there will be no compulsion on the insurer to underwrite your legal costs if you apply for insurance after the incident giving rise to the criminal proceedings.

The obvious question is, why bother, if States provide defence lawyers. It is a fact of life that the best legal advice and representation can cost more than the State affords a Defendant under a legal aid order. Professional indemnity insurance will underwrite the cost of such representation, although it will not indemnify the defendant for fines, because a contract of insurance cannot pay for criminal penalties as a matter of public policy.

CHOOSING A LAWYER

You need the best defence advice and representation, from the very start – in other words, from the point of arrest and questioning, even if that is beside the hospital bed. All too often, procedures are not followed, and damaging statements can be made unintentionally that the prosecution can exploit. This can occur even in regimes that one would consider above such things. In overseas jurisdictions, your Consul or Embassy will provide a list of lawyers, and the constituent names can be researched and consulted to find the best one for the job.

Carrying out some basic research along these lines is crucial because the quality of legal advice and representation depends upon the experience and expertise – and the independence – of the defence team. It is a dangerous thing to instruct a lawyer who has a basic knowledge of the criminal law but lacks experience in aviation criminalisation. This is a specialist field which demands an understanding of how to

DOI: 10.1201/9781003489696-8

apply the principles of negligence to the criminal process, and this will often involve searching for the best-experienced lawyer. If time is of the essence, instruct a criminal lawyer to protect your immediate interests, giving you some time and space to find the best lawyer with whom you will go to trial.

POINTS TO BEWARE

Lawyers sometimes have a weakness: they do not listen enough. It is important that you give them the full facts, upon which they will then consider the merits of the case in law against you to defeat the Prosecution's arguments. And remember that they are there to answer your questions as well as you should answer theirs.

At the end of a meeting, you must understand all that has been said, which leads to two things that you must do:

1. If there is anything which you do not understand, ask for clarification until you do understand.
2. At the end of the advice, ask the lawyer to summarise your options, and the merits of each. Get all of this in writing, so that you can mull it over and later refer back to it.

It is essential that you feel confident to put your trust in your lawyer, and this often means that the lawyer should be fluent in your language. But you do not need to have a friendly relationship with them. Just because they smile and are sympathetic has no bearing on their ability to get you out of jail.

Furthermore, remember that your lawyer must be a Rottweiler, and independence from the judicial process is essential. In some jurisdictions, your lawyer may have a personal agenda which is inconsistent with your interests. This may sometimes be revealed in their demeanour, or in their relations with the press; you should never permit them to give press interviews or statements.

PREPARING THE CASE

The relationship will start with an interview between you and your team, without the state being present or eavesdropping, when your team will listen to all you have to say and examine the evidence available.

The charge must be clear and unequivocal, so that the defence team can prepare the strongest arguments. At an early case management hearing, your lawyer will require the Prosecution to disclose all the evidence. It is important to ensure that this includes not just the evidence upon which they will rely, but also evidence which they have collected but which they do not intend to use.

In jurisdictions following English law, proof of *mens rea* and *actus reus* must be established beyond reasonable doubt. We have examined this in detail, so now is the time when you need to put your knowledge into practice. At the very least, focus on the essential elements of a crime: the *actus reus* contains all the elements in the definition of the crime except the defendant's mental element, which is defined by the *mens rea*, or guilty mind. If any element is missing, whatever the fault may be, it is not a crime. A strict liability offence does not require *mens rea*, but can be defeated by proof of due diligence, which need only be established on the balance of probabilities. A due diligence defence need not be established beyond reasonable doubt, but on the balance of probabilities. In other words, it must be more likely than not, based on 51 per cent to 49 per cent in the defendant's favour.

COLLECTION OF EVIDENCE

In whatever jurisdiction you find yourself, there is a key factor over which you have enormous control in the success of your case. This is the collection of your evidence. Here is a summary of how you will use your evidence in the case against you:

- The Prosecution bears the legal burden of persuading the court of the allegations beyond reasonable doubt. The task of your lawyer is to put sufficient doubt into the court's mind that the prosecution has managed this. It is dependent upon you, therefore, to produce such evidence in rebuttal – your defence team cannot produce it by magic.
- Your evidence must be delivered to the defence team without any colouring by you. At first sight to you, it may look bad, but it is for your team to manage the evidence and use it to discredit the prosecution's case. In any report or proof of evidence, only ever state the facts, never give an opinion or conclusion. If your lawyer asks you to explain, then you can elaborate.
- Eye-witness evidence will not be admitted by the court unless it goes some way to show, either directly or indirectly, that a fact in dispute did or did not exist. However, the court will not allow all evidence to be admitted. The Law of Evidence requires that certain kinds of relevant evidence cannot be considered, such as hearsay evidence that the witness did not see, say, hear or do themselves. If the prosecution offers such evidence, your lawyer can apply to have that evidence disallowed. The privilege against self-incrimination also rules out certain kinds of evidence.
- In addition to evidence gathered at the time, you will inevitably need some expert evidence, given by independent witnesses. Their duty is to help the court to achieve its overriding objective of dealing with the case justly, by giving objective, unbiased opinion on matters within their expertise. In the first instance, they will be given access to all the evidence, such as the documentation, VDR, photographs and statements. The expert witness will use this evidence to produce a report which will be submitted to the court, and will appear at trial to be cross-examined.

THE TRIAL

Under English law, in serious cases, you will be tried by a Jury, and in either-way cases, you have the right to elect to be tried by a Jury. The Judge is not permitted to exert any influence in directing the Jury to reach a verdict according to the evidence.

Appeals against conviction by a Jury are rarely successful unless the Appeal Court is satisfied that the Judge erred in law when directing them. You must, therefore, discuss with your defence team how fairly the Judge directed the Jury after they have retired to consider their verdict, so that the Judge's words are fresh in everybody's minds.

In other jurisdictions, the case may be put to an examining magistrate who will satisfy themselves that the prosecution evidence meets the minimum standards required under the criminal code. For example, the system in France involves the criminal court, which is a division of the regional court, but there are also six specialised maritime commercial courts, which try certain maritime offences. As general rule, the court is composed of three professional Judges, but there is also provision for specific offences to be dealt with by a Judge sitting alone.

THE RIGHT TO BAIL

We cannot leave this advice without a word about the right to bail. In many jurisdictions which follow international rules on human rights, the Defendant has the right to liberty until entering the dock at trial. There are grounds upon which the Prosecution can object, usually involving arguments that the Defendant will abscond and fail to appear at trial, or is likely to interfere with witnesses. A bail condition will invariably require the Defendant to reside at a suitably reliable address, such as a bail hostel, with which the Consul can assist with suggestions and of which the Court might approve. It will be essential for a senior representative of the hostel to attend court and answer any questions at the bail hearing; you may not get a second chance for bail, and it may be years before the case is listed for trial.

If the Defendant is held in custody contrary to the jurisdiction's criminal constitutional rules, they may appeal to a higher court. One solution, which, oddly, is seldom used, involves an application to the jurisdiction's constitutional process of Habeas Corpus, which may be invoked if the process of Court application has been exhausted. The writ of Habeas Corpus is intended to be a procedure to secure the release, as a matter of urgency, of a person who is unlawfully detained. The keyword is urgency, for it is considered as a safeguard to be invoked when there are no other remedies available. It will also ensure that illegal detention is not tolerated.

It is to be hoped that none of our readers will ever need the advice set out in this and previous articles – but it is well to be aware of it.

THE SECOND GOLDEN RULE

The second rule is *never, ever admit liability*. By doing so, your professional insurance policy is likely to be invalidated. At first sight, and certainly under pressure from the Prosecution team, your defence may appear difficult if not impossible; and they might propose a deal to mitigate your sentence. But it is for your defence team to discredit their case both as to the law and evidence. An admission of Guilt makes it very much more difficult for your lawyers to assist, whether the admission is made under pressure or otherwise.

Select Bibliography

ANSV, 2005, Final Report: Accident Involving ATR 72 Aircraft Registration Marks TST-LBB Ditching Off the Coast of Capo Gallo August 6th, 2005 (Tuninter Flight 1153).

Aaronson E and G Shaffer, 2021, Defining Crimes in a Global Age: Criminalization as a Transnational Legal Process, *Law & Social Inquiry* 46 (2) (May 2021) 455–486.

Air Transportation Safety Investigation Report A18P0115 (Abbotsford 2018).

Anon, 2000, Reforming the Law on Involuntary Manslaughter: The Government's Proposals, Home Office, May, 2000, CC NO77828.

Richardson P (ed), 2009 edition, *Archbold Criminal Pleading, Evidence and Practice*, Thomson Reuters (Legal) Limited, London.

Baker J, 2007, The Moral Limits of Criminalizing Remote Harms, *New Criminal Law Review* 11 (3) (2007) 371.

Bennun M, 1995, Prosecuting Professional Pilots in the United Kingdom after November Oscar: Reflections on the Law and Policy, *Journal of Air Law and Commerce* 61 (1995) 331.

Bercovitch J, 2004, International Mediation and Intractable Conflict, Beyond Intractability, in G Burgess and H Burgess (eds), *Conflict Research Consortium*, University of Colorado, Boulder.

Bhatnagar A and D Gajjar, 2024, How Is Artificial Intelligence Affecting Society? https://post.parliament.uk.

CAA, 2024, GM1 NCC.GEN.106 Pilot-in-Command Responsibilities and Authority.

CAA, 2020, Code of Practice for the Investigations and Enforcement Team, CAP 1422.

CAA, 2021, Security Management Systems (SeMS), A Guidance Note for Accountable Mangers.

Coleman J, 2002, *Markets, Morals and the Law*, Oxford University Press, Oxford.

Comisión De Investigación De Accidentes De Aviación Del Perú, Informe Final, 2023, Accidente De Aviación N° 008-2022, Ciaa-Accid-008-2022, Airbus A320n Cc-Bhb, Latam Airlines Perú SA (LATAM Perú Flight 2213).

Conseil Federal, 2020, Position et plan d'action du Conseil fédéral concernant la responsabilité des entreprises à l'égard de la société et de l'environnement.

CPS, 2018, The Code for Crown Prosecutors (currently 8th edition, October 2018). https://cps.gov.uk.

Demuijnck G, P Murphy and Philosophy Documentation Center, 2022, Retail Practitioners' Views vs. Ideal Theoretical Positions Concerning Ethical Business Practices with Garment Suppliers, *Business and Professional Ethics Journal* 41 (1) (2022) 59–92. doi:10.5840/bpej20211230114. ISSN 0277-2027. S2CID 24592700.

Department of Transport, 1987, mv Herald of Free Enterprise. Report of Court No. 8074. Formal Investigation.

Ericson R and K Carriere, 1994, The Fragmentation of Criminology, in D Nelken (ed), *The Futures of Criminology*, Sage, London.

Eurocontrol, 2018, Model for a Policy Regarding Criminal Investigation and Prosecution of Aviation and Railway Incidents and Accidents.

European Organisation for the Safety of Air Navigation, 2018, Model for a Policy Regarding Criminal, Investigation and Prosecution of Aviation and Railway Incidents and Accidents.

Federal Democratic Republic of Ethiopia Ministry of Transport and Logistics, 2022, Aircraft Accident Investigation Bureau Investigation Report Number AI-01/19, 23 December 2022 (Ethiopian Airlines Flight ET302).

Dekker S, 2012, *Just Culture: Balancing Safety and Accountability*, Taylor & Francis Group, Boca Raton, Fl.

Gold E (Professor), 2004, The Protection of Masters and Seafarers from Criminalization: Emerging Problems for the Shipping Industry, Centre for Maritime Law, University of Queensland, Brisbane Prepared for the 31st Annual Conference 'Navigating the Sea of Change' The Maritime Law Association of Australia and New Zealand Adelaide, South Australia 29 September–1 October 2004.

Hampton P, 2005, *Reducing Administrative Burdens: Effective Inspection and Enforcement*, H M Treasury, London.

Hudson B, 2003, *Justice in the Risk Society*, Sage, London.

ICAO, 1988, Airbus A300Bc, EP-IBU, Accident in the Vicinity of Qeshm Island, Islamic Republic of Iran on 3 July 1988. ICAO Circular 260-AN/154 (Iran Air Flight 655).

ICAO, 2004, Manual on the Regulation of International Air Transport (Doc 9626, Part 4).

ICAO, 2022, Event Involving Ryanair Flight Fr4978 in Belarus Airspace on 23 May 2021.

Komite Nasional Keselamatan Transportasi Republic of Indonesia, 2018, Aircraft Accident Investigation Report, Jakarta, November 2018 (Lion Air Flight 610).

Kun L, May 2008, Professional Responsibility: Politics, Culture, and Religion versus Science and Technology, *IEEE Open Journal of Engineering in Medicine and Biology* 27 (3) (2008) 95–99. doi:10.1109/MEMB.2008.918695. ISSN 0739-5175. PMID 18519190.

Law Commission, 2010, Consultation Paper No. 195 (Overview), Criminal Liability in Regulatory Contexts.

Lawrenson A and G Braithwaite, 2018, Regulation or Criminalisation: What Determines Legal Standards of Safety Culture in Commercial Aviation? *Safety Science* 102 (2018) 251–262.

Lindgren J, S Carlbaum, A Hult and C Segerholm, 10 December 2018, Professional Responsibility and Accountability?: Balancing Institutional Logics in the Enactment of New Regulations and Practices against Bullying and Degrading Treatment in Swedish Schools, *Nordic Studies in Education* 38 (4) (2018) 368–385. doi:10.18261/issn.1891-2018-04-06. ISSN 1891-5949. S2CID 58295321.

Lowenfield A, 1989, Looking Back and Looking Ahead, *American Journal of International Law* 83 (2) (1989).

Macrory, 2006, *Regulatory Justice: Sanctioning in a post-Hampton World Consultation Document*, Better Regulation Executive, London.

Michalowski R, 1985, *Order, Law and Crime: An Introduction to Criminology*, Random House, New York.

Michaelides-Mateou S and A Mateou, *Flying in the Face of Criminalization: The Safety Implications of Prosecuting Aviation Professionals for Accidents*, Routledge, Abingdon.

Mukherjee P, 2006, Criminalisation and Unfair Treatment: The seafarer's Perspective, *Journal of International Maritime Law* 12 (2006) 325–336.

NTSB Aviation Investigation Final Report accident number: CHI97FA084 (Newton, Wisconsin 1997).

NTSB, 2007, Attempted Takeoff from Wrong Runway, Comair Flight 5191, Bombardier CL-600-2B19, N431CA, Lexington, Kentucky, 27 August 2006, Aircraft Accident Report NTSB/AAR-07/05, Washington, DC.

Nardin T and D Mapel (eds), 1992, *Traditions of International Ethics*, Cambridge University Press, Cambridge.

Nemsick R and S Passeri, 2012, Criminalizing Aviation: Placing Blame before Safety, accessed via https://apps.americanbar.org/litigation/committees/masstorts/articles/winter2012-criminalizing-aviation-blame-safety.html.

Ormerod D, 2008, *Smith & Hogan Criminal Law*, Oxford University Press, Oxford.

Powell J and J Paterson, 2004, *Cicero The Advocate*, Oxford University Press, Oxford.
Reiner R, 2016, *Crime, The Mystery of the Common-Sense Concept*, Polity Press, Cambridge.
Ruitenberg B, 2002, *Court Case against Dutch Air Traffic Controllers*, LNVL, Amsterdam.
Solbrekke TD and T Englund, November 2011, Bringing Professional Responsibility Back In, *Studies in Higher Education* 36 (7) 847–861. doi:10.1080/03075079.2010.482205. ISSN 0307-5079. S2CID 144219656.
Tadros V, 2006, The Homicide Ladder (2006) 69(4) MLR 601–618.
Transportation Safety Board of Canada, Air Transportation Safety Investigation Report A19C0145, 2019 (Sachigo Lake).
Trögeler M, 2010, *Criminalisation of Aviation Accidents: Creation of a Just Culture: Balancing Aviation Safety and the Proper Administration of Justice*, Leiden University Law School, Leiden.
UK Attorney-General's Office, 2011, *The Dispute Resolution Commitment*, Ministry of Justice, London.
UK Parliament, 2023, House of Commons Library Briefing, Conflict in Ukraine: A Timeline (2014 – eve of 2022 invasion).
US Department of Justice Criminal Division, The Boeing Company Deferred Prosecution Agreement (DPA), 2024, Court Docket No.: 4:21-CR-005-O (N.D. Texas).
Wygle B, 1978, Performance Analysis of the Olympic Airways Takeoff at Athens on August 9, 1978 with an Engine Failure at Rotation, Boeing (Olympic Airways Flight 411).

TABLE OF CASES

Akcine Bendrove Bankas Snoras v Antonov and another [2013] EWHC 131 (Comm).
Andrews v DPP [1937] 2 All ER 552.
Bowater Steamship Co Ltd v Patterson and Others, The Gladys Bowater [1962] 303 F.2d 369 USCA.
Camarata Property Inc v Credit Suisse Securities (Europe) Ltd [2011] EWHC 479 (Comm).
Caparo Industries plc v Dickman [1990] 2 AC 605; [1990] [1990] UKHL 2.
Donoghue (or McAlister) v Stevenson, [1932] All ER Rep 1; [1932] AC 562.
Eweida v United Kingdom [2013] ECHR 37
Grimshaw v. Ford Motor Co (1981) 119 Cal.App.3d 757, 810.
HL Bolton (Engineering) Ltd v TJ Graham and Sons [1957] 1 QB 15.
Haughton v Smith (1975) AC 467.
Honey Maria Rose v R [2017] EWCA Crim 1168.
In Re Air Crash at Lexington, KY, 501 F.Supp.2d 902 (E.D. KY 2007).
In Re Air Crash at Lexington, KY, Aug. 27, 2006, 2008 WL 1909007 (E.D. KY 25 April 2008).
In Re Air Crash at Lexington, Kentucky, Civil Action (Master File) No. 5:06-Cv-316 – Ksf (E.D. Ky. 2 Febuary 2011).
Kingdom of the Netherlands v Russian Federation, ITLOS Case No. 22, Order 2013.
Lister and Others (AP) v Hesley Hall Limited [2001] UKHL 22.
Magill v Porter [2002] 2 AC 357.
Mangouras v Spain (12050/04).

Marc Rich & Co AG and Others v Bishop Rock Marine Co Ltd and Others [1995] 3 All ER 307; [1995] UKHL 4; [1996] 1 AC 211; [1995] CLC 934; [1995] 2 LLR 299; [1996] ECC 120; [1995] 3 WLR 227; [1995] 2 Lloyd's Rep 299.
Northern Securities Co v United States 193 US 197 (1904).
Ocean Prefect Shipping Limited v. Dampskibsselskabet Norden AS, the Ocean Prefect [2019] EWHC 3368 (Comm).
Tesco Supermarkets Ltd v Nattrass [1972] AC 153 [HL].
R v Adomako [1995] 1 AC 171.
R v Bateman (1925) 19 Cr App R 8.
R. v CAV Aerospace Ltd 2015, Central Criminal Court (unreported).
R v Caldwell [1981] 1 All ER 961.
R v Cheshire [1991] 1 WLR 844.
R v Deller [1952] 36 Cr App Rep 184.
R v Evans [2009] EWCA Crim 650.
R v Fulling [1987] 2 All ER 65.
R v G and Another [2003] UKHL 50; [2004] 1 AC 1034.
R v HM Coroner for Inner London, ex parte Douglas-Williams [1998] 1 All ER 344.
R v Henderson, 2021, Cardiff Crown Court (unreported).
R v K [2001] UKHL 41.
R v Lawrence [1981] 1 All ER 974.
R v Misra and Srivastava [2004] EWCA Crim 2375.
R v Moloney [1985] 1 All ER 1025.
R v Murgatroyd, 2017, Manchester Crown Court (unreported).
R v Nedrick [1986] 3 All ER 1.
R v O'Connor [1997] Crim LR 16 CA.
R v P & O European Ferries (Dover) Ltd [1991] 93 Cr App R 72.
R v Prince (1875) LR 2 CCR 154.
R v Rudling [2016] EWCA Crim 741.
R v Secretary of State, ex parte Kirkstall Valley [1996] 3 ALL ER 304.
R v Sellu [2016] EWCA Crim 1716.
R v Sussex Justices, Ex parte McCarthy [1924] 1 KB 256, [1923] All ER Rep 233.
R v William Glen Stewart [1991] Isleworth Crown Court (unreported).
R v Williams (Gladstone) [1987] 3 All ER 411.
Re Medicaments and Related Classes of Goods (No. 2) [2001] 1 WLR 700.
Scott Hoyle v Julia Mary Rogers and Nicola Lucinda Rogers and Secretary of State for Transport (1st Intervener) and International Air Transport Association (2nd Intervener) [2014] 3 WLR 148; [2014] CP Rep 30; [2014] EWCA Civ 257; [2014] 3 All ER 550.
Sweet v Parsley [1970] AC 132, at p. 148.
The Argonaftis [1989] 2 Ll R 487.
United States of America v Wolfgang Schröder [2007] United States District Court, Alabama Southern District (unreported).
United States v The Boeing Company (currently unreported).
Williams and Another v Natural Life Health Foods Ltd [1998] 1 WLR 830.

TABLE OF CASE STUDIES

Abbotsford 2018	Chapter 3	(Canada)
Amsterdam 1998	Chapter 2	(Netherlands)
Brown, Roderick	Chapter 1	(UK)
Captain Chawla of the *Hebei Spirit*	Chapter 6	(South Korea)
Captain Mangouras of the *Prestige*	Chapter 3	(Spain and ECHR)
Captain Schröder of the *Zim Mexico III*	Chapter 3	(USA)
Captain Stewart and November Oscar	Chapter 2	(UK)
CAV Aerospace	Chapter 5	(UK)
China Eastern Flight MU5643	Chapter 2	(China)
Ethiopian Airlines Flight ET302	Chapter 1	(Ethiopia)
Forkner, Mark	Chapter 1	(USA)
Grimshaw v Ford	Chapters 2 and 5	(USA)
Hoyle v Rogers	Chapter 3	(UK)
Iran Air Flight 655	Chapter 5	(Iran)
LATAM Perú Flight 2213	Chapters 2 and 6	(Peru)
Lexington 2006	Chapter 2	(USA)
Lion Air Flight 610	Chapter 1	(Indonesia)
Malaysia Airlines Flight 17	Chapter 5	(Ukraine)
Mount Erebus 1979	Chapter 4	(Antarctica)
Murgatroyd, Robert	Chapter 5	(UK)
Newton, Wisconsin 1997	Chapter 1	(USA)
Olympic Airways Flight 411	Chapter 5	(Greece)
Paris 2000	Chapter 3	(France)
R v Henderson	Chapter 1	(UK)
Ryanair Flight 4978	Chapters 3 and 5	(Belarus)
Sachigo Lake 2019	Chapter 4	(Canada)
Tuninter Flight 1153	Chapter 2	(Italy)
Turkish Airlines Flight 981	Chapter 2	(France)
Überlingen 2002	Chapter 2	(Switzerland-Germany)
United States v The Boeing Company	(See Forkner, Mark)	
ValuJuet DC-9 1996	Chapter 2	(USA)
Zurich 2012	Chapter 2	(Italy)

TABLE OF INTERNATIONAL CONVENTIONS AND EUROPEAN REGULATIONS

TABLE OF STATUTES

TABLE OF STATUTORY INSTRUMENTS

Index

FSC
www.fsc.org
MIX
Papier | Fördert gute Waldnutzung
FSC® C083411

Kolibri 360 GmbH
Ferdinand-Jühlke-Straße 7
99095 Erfurt, Deutschland
produktsicherheit@kolibri360.de